The Molecular Basis of
B-Cell Differentiation
and Function

NATO ASI Series

Advanced Science Institutes Series

A series presenting the results of activities sponsored by the NATO Science Committee, which aims at the dissemination of advanced scientific and technological knowledge, with a view to strengthening links between scientific communities.

The series is published by an international board of publishers in conjunction with the NATO Scientific Affairs Division

A	**Life Sciences**	Plenum Publishing Corporation
B	**Physics**	New York and London
C	**Mathematical and Physical Sciences**	D. Reidel Publishing Company Dordrecht, Boston, and Lancaster
D	**Behavioral and Social Sciences**	Martinus Nijhoff Publishers
E	**Engineering and Materials Sciences**	The Hague, Boston, Dordrecht, and Lancaster
F	**Computer and Systems Sciences**	Springer-Verlag
G	**Ecological Sciences**	Berlin, Heidelberg, New York, London,
H	**Cell Biology**	Paris, and Tokyo

Recent Volumes in this Series

Series A: Life Sciences

The Molecular Basis of B-Cell Differentiation and Function

Edited by

M. Ferrarini

Istituto Nazionale per la Ricerca sul Cancro
Genova, Italy

and

B. Pernis

College of Physicians and Surgeons
Columbia University
New York, New York

Plenum Press
New York and London
Published in cooperation with NATO Scientific Affairs Division

Proceedings of a NATO Advanced Study Institute on
the Molecular Basis of B-Cell Differentiation and Function,
held October 1–11, 1985,
in Santa Margherita Ligure, Italy

Library of Congress Cataloging in Publication Data

NATO Advanced Study Institute on the Molecular Basis of B-Cell Differentiation
and Function (1985: Santa Margherita Ligure, Italy)
    The molecular basis of B-cell differentiation and function.
    (NATO ASI series. Series A, Life sciences; v. 123)
    "Proceedings of a NATO Advanced Study Institute on the Molecular Basis of
B-Cell Differentiation and Function, held October 1–11, 1985, in Santa Margherita
Ligure, Italy"—T.p. verso.
    Includes bibliographies and index.
    1. B cells—Congresses. 2. Cell differentiation—Congresses. 3. Immunogenet-
ics—Congresses. I. Ferrarini, M. II. Pernis, Benvenuto. III. Title. IV. Series. [DNLM:
1. Cell Differentiation—congresses. 2. B Lymphocytes—congresses. WH 200
N279m 1985]
QR185.8.L9N38   1985   616.07'9   86-30596

ISBN 978-1-4684-7037-6   ISBN 978-1-4684-7035-2 (eBook)
DOI 10.1007/978-1-4684-7035-2

© 1986 Plenum Press, New York
Softcover reprint of the hardcover 1st edition 1986

A Division of Plenum Publishing Corporation
233 Spring Street, New York, N.Y. 10013

PREFACE

The progress in protein and nucleic acid chemistry together with improvements of the previously employed tissue culture techniques have led to the solution of problems such as that of the generation of antibody diversity or of the molecular structure of T and B cell membrane receptor for antigen which had challenged the past generations of immunologists. Thanks to this progress an impressive amount of knowledge has been accumulated on certain cell types that were relatively "mysterious" until recently. The B lymphocyte represents a typical example of such a cell.

With these considerations in mind, we have started to organize a NATO summer school on "The molecular basis of B cell differentation and function" that had the specific aim of bringing up to date a selected number of young investigators. During the course, held in Santa Margherita Ligure, October 1-11, 1985 it became apparent that rather than a formal school we had organized an informal series of Workshops where both the Faculty members and the participants were discussing their own data in addition to reviewing the general progress in the field. The deep committment of everyone to his research subject and the enthusiasm had caused this very successful change in the shape of the course. At the end we asked the members of the Faculty and the participants to write a summary of their data and of their point of view on selected subjects. Although not everyone has been able to contribute to this book, we believe that the papers collected provide a comprehensive view of the main topics discussed at the course and perhaps give an idea of the atmosphere created at the course itself.

We would like to thank all those who have contributed to the course and the book. In particular, NATO has provided the necessary financial support and Dr. C. Sinclair, Director of Advanced Scientific Institute program, has helped us a great deal in solving a number of problems.

Dr. L. Santi, Director of Istituto Nazionale per la Ricerca sul Cancro, Genova, beside financial support, has provided the use of the facilities of the Scuola Superiore di Oncologia e Scienze Biomediche in Santa Margherita Ligure and the precious help of their staff. National

v

Institute of Allergy and Infectious Disease, USA and National Science Foundation, USA have also co-sponsored the course. Dr. M.D. Cooper (Birmingham, ALA) and Dr. M.F. Greaves (London) have served in the advisory board and their advice and help has been most precious Dr. A.S. Fauci (Bethesda, MD) has also helped us both with his advice and in finding extra financial support. Simonetta Zupo and Vito Pistoia have contributed to give the shape of a book to what was a collection of papers.

M. Ferrarini
B. Pernis

CONTENTS

REGULATION OF THE ASSEMBLY AND EXPRESSION OF IMMUNOGLOBULIN GENES:

VARIABLE REGION ASSEMBLY AND HEAVY CHAIN CLASS SWITCHING

R. A. DePinho, G. D. Yancopoulos, T. K. Blackwell,
M. G. Reth, K. Kruger, S. G. Lutzker, and F. W. Alt

College of Physicians and Surgeons of Columbia University
701 West 168th Street
New York, NY 10032

INTRODUCTION

To elucidate further the general mechanisms which control the assembly and expression of immunoglobulin variable region gene segments and heavy-chain class switching, we have exploited Abelson murine leukemia virus (A-MuLV) transformed pre-B cells as a model system. The Abelson virus is a replication defective retrovirus which is unique in its capacity to transform very early B-lineage cells in vitro (1). A-MuLV transformants generated from murine fetal liver or adult bone marrow have provided both a static representation of various pre-B cell differentiation stages, as well as a dynamic view of pre-B cell immunodifferentiation events (2,3). Our analyses of these lines have provided a variety of new and sometimes surprising findings regarding the early stages of B cell differentiation; and importantly, all of these novel aspects first observed in A-MuLV transformants were found to reflect events that occur early in normal B lineage cells in vitro. Furthermore, we have utilized gene transfer technology to gain additional insight into the molecular events involved with the regulation of immunoglobilin and T cell receptor variable region (V) gene assembly (3).

THE ANTIBODY MOLECULE: STRUCTURE AND FUNCTION

The prototypical IgG molecule is composed of two identical light chains (LC) and two identical heavy chains (HC) which are covalently linked by disulfide bonds (4,5) (Figure 1). Through extensive peptide sequence comparisons of both the amino-terminal variable region and the carboxy-terminal constant region of myeloma proteins and monoclonal antibodies, it has been established that different antibodies differ from each other primarily in their variable regions and, within a particular class or subclass, are identical in their constant regions (4-7). The variable region of heavy and light chains can be further subdivided into regions of relatively conserved amino acid sequence (framework regions) and regions of highly variable amino acid sequence referred to as complementarity determining regions (CDRs) (7). The CDRs are a major site of antibody diversity and they are thought to be areas of the antibody directly involved in antigen contact (8). The remainder of the antibody is composed of a region of constant amino acid sequence, which may be one of several classes or subclasses for the HC:$\mu, \delta, \gamma 3, \gamma 1,$

1

$\gamma 2b, \gamma 2a, \varepsilon,$ and α; and may be one of two families for the LC: Kappa (K) and Lambda (λ). The constant region of the HC has been implicated in a variety of effector functions, among them complement fixation, binding to Fc receptors on phagocytes, and placental transfer (9).

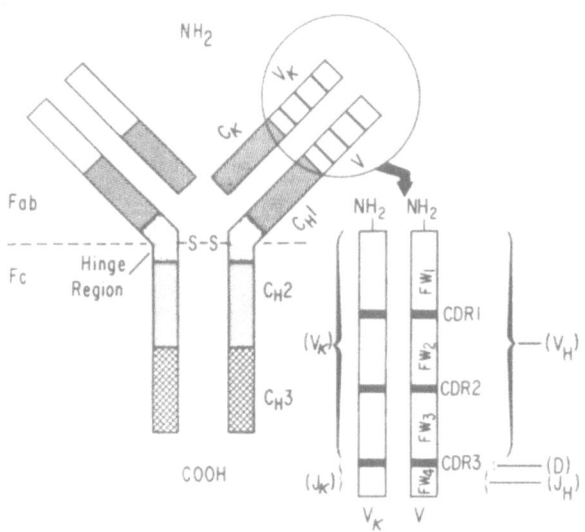

Figure 1. A prototypical IgG,K antibody molecule and the germline gene segments which encode for various portions of the variable region.

The Molecular Basis of Antibody Diversity. Variable Region Gene Structure and Assembly

The variable regions of an immunoglobulin HC and LC genes are encoded by multiple genetic elements (for review see ref. 8). Early in the B cell differentiation pathway, these segments are assembled to form complete variable region (V) genes (Figure 2). For the HC V gene individual members from three separate clusters of gene segments are assembled (10-12). These include the V(H) segment which encodes the bulk of the gene; the D (diversity) segment which encodes for most of CDR3; and the J(H) (joining) segment which encodes for the fourth framework region (Fig. 1 and 2). In the mouse there are 4 J(H) segments which lie approximately 7 kb upstream from the first constant region

2

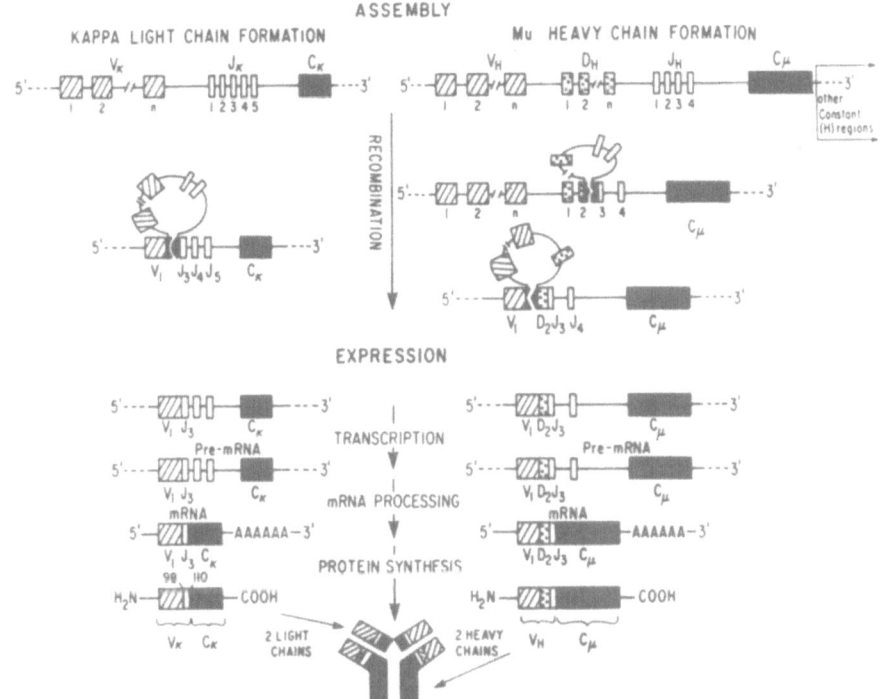

Figure 2. Organization, ordered assembly and expression
of a productive HC and LC variable region rearrangement.

gene, Cμ (11). The 12 known D segments lie in the 80kb of DNA just
upstream from the cluster of J(H) segments (13,14).   At an unknown
distance 5' to the D cluster there are 100 - 1000 V(H) gene segments
that are organized into approximately 7 families based on nucleic acid
sequence homology (15-19).   The LC variable region is encoded only by
V(L) and J(L) segments (8). The Kappa LC V gene is assembled from any
one of the 100 - 300 V(K) gene segments each of which is capable of
encoding the first and second CDR and their surrounding framework
residues (8,20,21).   The remainder of the LC V gene is encoded by one of
5 J(K) gene segments each of which can encode a portion of the third CDR
and the entire fourth framework region (22,23).   Still further
downstream to the J(K) cluster is a single K constant region gene.   The
λ locus is organized in a somewhat different fashion with only two V(λ)
genes, each of which is followed downstream by two J(λ) - C(λ) units
(24).

     Two distinct recombinational events are involved in the assembly of
the HC V gene; a D segment is joined to a J(H) segment and subsequently

a V(H) segment is joined to the pre-existing DJ(H) complex (25). Assembly of a LC V gene involves a single event, the joining of a V(L) to a J(L) (9). The actual joining of these gene segments appears to occur by a two-step, non-reciprocal recombination process (26) which is targeted by a set of conserved recombination recognition sequences, consisting of a palindromic heptamer and a characteristic nonamer, separated by a spacer of either 12 or 23 base pairs (bp) (10,11). A complete recognition sequence containing a 23 bp spacer directly abuts the 5' side of each J(H) segment and the 3' side of each V(H) segment. Both the 5' and 3' side of each D segment is flanked by recognition sequences containing 12 bp spacers. Apparently, recombination events only take place between segments if one of the segments has a recognition sequence with a 12 bp spacer and the other has a recognition sequence with a 23 bp spacer. In this way, the length of the spacers can direct the recombination process between various segments. Thus, a V(H) and J(H) always recombine with the 5' and 3' side of the D segment, respectively. This has been referred to as the 12/23 rule (10,11). Consistent with this rule, the V and J segments of the LC, which must be joined directly, are flanked by recognition elements containing 12 or 23 bp spacers. Given the extensive number of gene segments contained in the immunoglobulin locus, the random combinatiorial association between each these genetic elements could account for an enormous antibody repertoire; however, detailed analysis of rearranged segments reveals that additional mechanisms serve to further increase antibody diversity. In particular, an examination of recombination joints between various segments has shown that the joining process is imprecise, in that nucleotides can be deleted from the potential coding regions of the HC or LC V gene segments and nucleotides can be inserted at the junctions of the HC V gene segments (26). The inserted nucleotides, referred to as N regions, are thought to be added by the enzyme terminal deoxynucleotydl transferase (TdT) (26,27). N regions have not been observed at the junctions of LC V gene segments. Additional evidence for various proposed aspects of this recombination process including the role of TdT in the addition of N regions has been provided by studies involving the introduction of recombination substrates into certain A-MuLV transformants (28,29,Blackwell et al., submitted; see below).

Expression of Immunogloublin Genes

Although each germline V(H) gene segment is preceded by its own transcriptional promoter region, in mature Ig-producing B cells, only the promoter of the rearranged V(H) gene segment is active (30); whereas, those promoters upstream of unrearranged V(H) segments are transcriptionally silent. The exclusive transcriptional activity of the rearranged V(H) promoter in the mature B cell is thought to be dependent upon its close promimity to a tissue-specific enhancer, which is located in the intron between the J(H) and C exons (31-33). While the enhancer has been shown to be an important element in Ig expression, additional investigations indicate that other sequences are also necessary for high level, tissue-specific expression; in particular, sequences surrounding the V(H) promoter (34,35). Initial evidence for tissue or stage-specific regulatory elements associated with the V(H) segment was provided by the finding that germline V(H) gene segments are capable of high level expression in pre-B cells that are actively undergoing V(H) to DJ(H) joining (30). Germline V(H) transcripts initiate from the normal V(H) promoter, are properly spliced, polyadenylated, and appear in the cytoplasm. High level expression has only been observed thus far for members of the J558 V(H) region family. It is not known whether the V(H) transcripts are translated into V(H) polypeptides and what, if any, biological function such proteins would possess (for review see refs. 2 and 3).

In addition to germline V(H) expression, A-MuLV transformed pre-B cell lines and other B lineage cells produce truncated $C\mu$-positive HC transcripts that originate from DJ(H) rearrangements and thus lack a classical V(H) component (36). Transcription of $D\mu$ mRNAs initiates from promoters that are located upstream from most or all D segments. These $D\mu$ transcripts are translated if a translational initiation codon upstream from the D segment is in the proper reading frame with the joined J(H) segment. About 40 to 60 amino acids are provided by the D segment and, strikingly, most $D\mu$ proteins contain a hydrophobic leader-like sequence. Given that the promoter, translation start site codon and encoded leader-like sequences seem to be conserved across species, a function for the expression of DJ(H)-μ proteins seems probable.

REGULATION OF VARIABLE REGION GENE ASSEMBLY

Unique mechanisms appear to have evolved to ensure the generation of mono-specific B cells, a fundamental prerequisite to an immune system whose specificity is based upon antigen-driven clonal selection. In particular, the ordered rearrangement events characteristic of B cell differentiation appear to be involved in a critical regulatory phenomenon in V gene assembly, the phenomenon of "allelic exclusion" (37,38). Allelic exclusion refers to the observation that any given B-cell assembles a productive rearrangement at just one of its two HC alleles and a productive rearrangement at just one of its several LC alleles (39). Thus, although the potential exists for a B cell to assemble and express more than one of its Ig HC and LC loci, allelic exclusion limits any given B cell clone to the production of a single heavy and light chain which associate to form the surface Ig receptor. Before we discuss the mechanisms which regulate HC and LC V gene assembly, we must consider the relationship between the assembly and the functional expression of immunoglobulin genes.

Rearrangement Imprecision and Allelic Exclusion

Rearrangements can be either "productive" (functional) or "nonproductive" (nonfunctional). A productive rearrangement results when a V gene segment is joined to a J segment in the correct, in-phase translational reading frame and subsequent expression leads to a complete protein product. Nonproductive rearrangements do not encode for complete Ig proteins (40,41). An example of such a rearrangement is when a V gene segment, due to the random insertion and deletion phenomena mentioned above, is joined to the J gene segment in a different translational reading frame; although the resulting rearrangement may be expressed in the form of a complete V(D)JCμ mRNA, this RNA could not be translated into a complete protein product (40,42). Assuming the insertion and deletion of bases at the junctions of V gene segments occurs randomly with respect to reading frame, approximately 70% of the joins should occur out-of-phase. This number, however, is probably somewhat of an underestimate because several other mechanisms exist to generate non-productive or non-functional rearrangements (for review see 2).

The regulation of HC allelic exclusion and the onset of LC V gene assembly has been proposed to be effected by the μ protein from a productive rearrangement which, in some way, feeds back to prevent further rearrangement of a V(H) segment to a DJ(H) complex and also signals V(L) to J(L) rearrangement (25,43,44). The μ HC protein thus appears to provide two important signals in the regulated model of allelic exclusion: a negative signal which terminates additional V(H) rearrangement on the other allele and a positive signal which leads to V(L) to J(L) rearrangement. Light chain allelic exclusion apears to be regulated by testing each V(L)J(L) rearrangement for the produciton of a LC which can functionally interact with the pre-existing HC in the cell; the formation of a complete Ig apparently leads to cessation of further LC gene rearrangement (40).

Support for the "regulated" model of variable region gene assembly has come from a variety of systems and studies (for review, see ref. 2); however, many of these regulatory features are exhibited by certain A-MuLV transformed cell lines which immunodifferentiate during growth in culture. The best characterized example is that of the 300-19 cell line (43). The original transformant (300-19P) represents the earliest identified stage of pre-B cell differentiation in that it has DJ(H) rearrangements on both chromosomes and no LC variable region gene segment rearrangements. When grown in culture, 300-19P spontaneously appends V(H) gene segments to its pre-existing DJ(H) complexes. During this stage of rearrangement, 300-19P produces both germline V(H) transcripts and Dμ proteins. Although it was proposed that production of complete μ chains signal a cessation of V(H) to DJ(H) joining, the production of Dμ proteins do not prevent rearrangement from occurring, apparently because their production level is too low. Interestingly, one isolate of 300-19 which produces abnormally high levels of Dμ protein does not appear to assemble complete HC V genes, but instead goes on to assemble V(K) genes (43). This finding led to the suggestion that the constant portion of the HC may be sufficient to mediate the regulatory signal. Preliminary data suggest that when the 300-19 line forms a productive rearrangement on its first attempt, the second allele is frozen in the DJ(H) configuration. When the first V(H) to DJ(H) rearrangement is non-functional, a V(H) gene segment is then appended to the second DJ(H) allele. When the second rearrangement is non-functional, the line does not appear to undergo any additional rearrangement events; it becomes a dead-end null cell. If either the first or the second rearrangement event is productive (generates a protein), most subclones go on to rearrange their K LC genes. Thus, 300-19 subclones which had undergone productive HC rearrangements assembled their K chain V genes at a rapid rate; subclones which had generated two nonproductive HC rearrangements failed to assemble their V(L) loci. LC gene assembly also appears to be regulated temporarily in that V(K) genes are assembled before V(λ) genes; this property is clearly evident in the 300-19 line. The nature of the potential mechanisms which effect the ordered assembly of V(K) and subsequently V(λ) genes has not been elucidated. As mentioned above, assembly of the LC V genes is also an allelically excluded event; at this point, it is not clear whether the 300-19 or any A-MuLV transformants manifest allelic exclusion at the level of V(L) assembly.

The Accessibility Model for the Control of V Gene Assembly

Our studies have clearly indicated that tissue-specific, developmental stage-specific, and allelically excluded assembly of Ig HC and LC V genes are very highly regulated processes. We have proposed that the assembly of the V genes of both the Ig HC and LC and T cell receptors may be regulated at the level of the accessibility of the component gene segments to a common recombinase (3,29). According to this model, the μ HC could mediate its dual regulatory function by signaling for the "closing" of germline V(H) gene segments (making them inaccessable to recombinase) and the "opening" of V(K) gene segments (making them accessible to recombinase) (43). Likewise, this model suggests that Ig V gene segments are completely assembled in B cells and

that T cell receptor V gene segments are assembled in T cells because of
the relative accessibility of the different types of V gene segments in
the two lineages to a "constitutive" recombinase. Among the initial
observations that argued for a common recombinase was the finding that
both the Ig and T cell receptor V gene segments are flanked by
recombination recognition sequences which are very similar with no
consistent differences between them (reviewed in 45). Initial evidence
supporting the accessibility model included the observations that
unrearranged Ig and TCR V gene segments appear to become
transcriptionally active just at or prior to the cell stage in which
they are normally rearranged; thus, such transcription could reflect
increased accessibility of these gene segments to recombinational as
well as to transcriptional machinery (3,30,Blackwell et al., submitted).

    More recently, we have employed gene transfer studies to test the
postulates of the accessibility model. The experiments involve the gene
transfer of constructed recombination plasmids containing various gene
segments, such as V, D or J gene segments derived from either Ig or T
cell receptor genes, into pre-B cells that are actively rearranging
their endogenous Ig genes (28,29,46, Blackwell et al., submitted).
These constructs which also contain a linked herpes simplex virus
thymidine kinase (tk) gene are introduced into tk-minus derivatives of
"recombinase-positive" fetal liver transformants by unlinked
co-transformation with pSV2-neo (T cell receptor constructs were
selected for tk expression in HAT media). Clones containing the
integrated construct are isolated by selection for expression of the
pSV2-neo gene in the presence of a neomycin analog, G418. The
G418-resistant subclones can then be placed in conditions which select
for the expression of the neighboring tk gene. For the most part cells
in non-tk-selective conditions do not express the introduced tk gene;
and significantly, such cells show little evidence of recombination
substrate gene rearrangement. Strikingly, when subclones were selected
for the expression of the flanking tk gene, nearly all of these
subclones actively rearrange the gene segments within the integrated
construct (46,Blackwell et al., submitted). Thus, it appears that the
frequency of D to J(H) joining, within a recombination substrate
introduced into a rearranging pre-B cell line, could be regulated by
controlling the expression of an adjacent selectable marker gene; an
effect perhaps mediated through a local alteration of chromatin
structure or by transcription per se (46). To test the common
recombinase hypothesis, we have introduced either lambda LC or T cell
receptor V gene segments into pre-B cell lines which were only
recombining endogenous HC V gene segments (29, Blackwell et al.,
submitted). Although the endogenous lambda LC or T cell receptor V gene
segments did not rearrange in this line, the corresponding introduced
segments were assembled at a high rate when in an accessible
configuration (i.e., when the linked marker gene was expressed). For
both the introduced lambda LC and T cell receptor V gene segments we
have provided a structural correlate of accessibility, i.e., the
actively rearranged introduced segments were in a DNAase-sensitive state
while the corresponding endogenous segments were not.

Preferential V Segment Utilization Based on Chromosomal Position

    A-MuLV which actively appended V(H) to DJ(H) segments in culture
provided a unique system to examine the initial V(H) usage pattern in
the absence of obvious antigenic or immunoregulatory selective forces.
An examination of the V(H) segments used by such A-MuLV transformants
have revealed that the most J(H)-proximal V(H) gene segments are
preferentially used to form V(H)DJ(H) rearrangements in pre-B cells

lines (47,48). To elucidate the development of the V(H) repertoire in the animal, we determined the relative utilization of different V(H) families in normal B cell populations of the developing mouse (Yancopoulos, et al., submitted). These studies indicated a clear position-dependent bias for V(H) rearrangement extending across the entire V(H) locus during early fetal development, suggesting that the early fetal repertoire may simply be a reflection of the position dependent rate at which particular V gene segments are rearranged. In contrast, the V(H) usage pattern seen in the mature B cells of adult spleen reveals a more randomized, non-position-dependent repertoire. Thus, seletion by either antigen, accessory immune cells, or other mechanisms appears to play a major role in establishing the more balanced adult V(H) repertoire. This interpretation is supported by preliminary evidence which indicates that mice with a severe combined immunodeficiency disease do not generate functional HC protein; and correspondingly that the V(H) repertoire of the "defective" HC mRNA found in the adult spleens of these mice resembles that of early fetal development. Interestingly, a repertoire reminiscent of that of the normal fetus appears to be maintained in autoantibodies suggesting the possibility that these anti-self antibodies are selected at an early stage in development or from a B cell sub-population which maintains a more "fetal"-like repertoire (49). The direct relationship between chromosomal position and rearrangement utilization frequency of V(H) segments is consistent with the possibility that recombinase acts via a one-dimensional tracking mechanism in which the enzyme proceeds from the DJ(H) complex or other entry sites to the V(H) locus.

HEAVY CHAIN CLASS SWITCHING IN PRE-B CELLS

During the differentiation of B lymphocytes, functional heavy and light Ig chains appear first in the form of IgM membrane receptors on the immature B lymphocyte. Upon interaction with cognate antigens or certain mitogens, a B cell is induced to proliferate and mature, ultimately leading to the generation of the terminally differentiated plasma cell which secretes Ig with the same variable region specificity as was on the surface of its progenitor B cell clone (reviewed in 50,51). Previous studies have demonstrated that the molecular mechanism employed in the maintenance of V(H) gene specificity in cells undergoing the transition from membrane to secreted Ig production involve differential processing of appropriate mRNA precursors (52). In addition, during the maturation of the B cell to the plasma cell, the class of Ig produced by the progeny may change from IgM to a different class (i.e. IgG or IgA), while maintaining the same V region specificity (53). This phenomenon, termed HC class switching, allows a single clone of B cells to generate progeny which maintain the same antigen binding specificity V(H)DJ(H) linked to different C(H) effector functions. The C(H) locus has been well characterized; it spans about 200 kb and is organized as detailed in Figure 3a (54). Mechanistically, it appears that the class switch process is often effected by a recombination event mediated by a somewhat conserved set of repeated sequences (S regions) which span several kb 5' to each C(H) gene (reviewed in 50). In its most simple form, such a switching mechanism involves the direct joining of the switch region upstream of the C_μ gene to the switch region of the downstream C(H) region to be expressed (Figure 3b) accompanied by the deletion of all of the intervening DNA including the C_μ and any other C(H) genes; this mechanism is referred to as the classical recombination/deletion mechanism. Sequence comparisons of the various S regions with S_μ indicates that Sℓ is most homologous while Sγ2b is the least homologous (51). Detailed studies of the sequence organization surrounding the recombination joints have led some investigators to propose that the site of recombination may be directed by specific

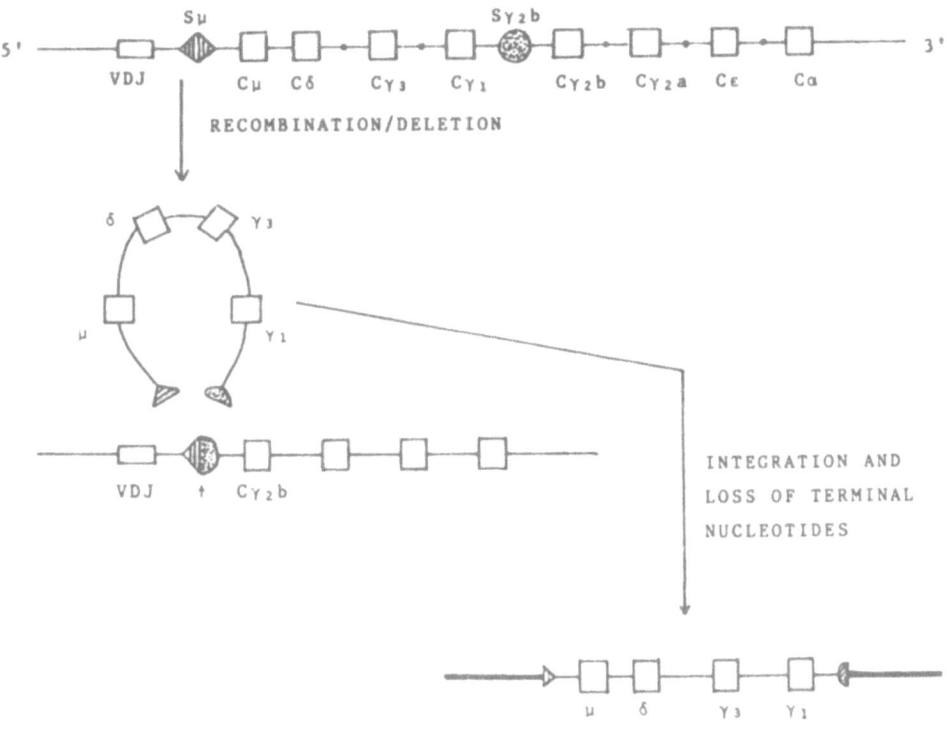

18-81 Pre-B Cell γ2b Class Switch:
RECOMBINATION/DELETION/REINTEGRATION

Figure 3. (A) Organization of the constant region locus.
(B) Recombination/deletion leading to a μ to γ2b class
switch. (C) Re-integration of the excised intervening
region into uncharacterized sequences.

recognition sequences (55) or by potential stemloop structures (56). As
these examples constitute a minority of the cases examined, it seems
more likely that the overall organization of the S region serves as a
general recognition site for proteins involved in the rearrangement
process and that the recombinase may cut and ligate at multiple (perhaps
any) sites along the S region or its flanking sequences.

A number of independent A-MuLV transformed pre-B cells have been
demonstrated to undergo switching in culture (57,58) and thus provide a
model system, analogous to that described above for the study of V gene
assembly, to study the molecular mechanisms involved in HC class
switching. In this regard, we have examined the molecular basis of HC
class switch events in several independently isolated transformants
which were derived from different mouse strains. Our studies have
demonstrated that switching in these lines occurs by the classical
recombination/deletion mechanism and usually involves switching from μ
to γ2b, suggesting that the propensity to switch to this particular
isotype might be a general property of pre-B cells (58). In addition,

our studies have revealed several surprising mechanistic aspects of the switch process and may provide some novel insights into the general mechanisms by which switching is controlled.

Mechanistic Aspects of Class-Switching in A-MuLV Transformants

HC class switching is a high frequency event in some isolates of the 18-8 pre-B transformant (59,60). We previously demonstrated that a subclone of this line switched from μ to γ2b expression by the classical recombination/deletion mechanism (58). In this line, the expressed V(H)DJ(H) region and the Cγ2b constant region were juxtaposed by a recombination event which linked the highly repetitive portion of the Sμ and Sγ2b regions and resulted in the loss of the Cμ gene from the intervening region. Despite the recombination/deletion mechanism of switching, the γ2b-producing line retained two copies of the Cμ gene and two copies of the sequence just 5' to the Sγ2b recombination point. These sequences manifested as novel fragments which hybridized with specific 3' Sμ and a 5' S 2b probes. These "retained" sequences were molecularly cloned and analyzed in detail; the nucleotide sequence of the novel junctions within these clones demonstrated that the 3'Sμ clone contains 3' elements of the repetitive Sμ region joined to uncharacterized sequences and that the 5' Sγ2b clone contained a region of the classical 49 bp Sγ2b repeat region again joined to uncharacterized sequences (figure 4). The location of the novel joints was close to (within 400 bp) but not at the switch recombination point. We had previously proposed several alternative models for the retention of intervening sequences in these lines. The current data is most consistent with a model involving re-insertion into a chromosome of linear DNA sequences which were excised from between the switch recombination points (figure 3c). The lack of perfect reciprosity between the joints in the retained sequences and the switch recombination point probably results from deletion of sequences near the ends of the re-integrated linears (DePinho et al., in preparation).

A detailed analysis of a classical μ to γ2b class switch event in an independent A-MuLV transformant, 300-18 (PD6), has revealed that class switch events can also involve complex recombination processes (Yancopoulos, DePinho, Zimmerman and Alt, submitted). In this case, a recombination-deletion event involving inversion of a segment of the Sγ2b region had taken place. Again, a detailed elucidation of the recombination mechanism was determined by direct nucleotide sequence analysis of the involved regions. In this line, the recombination joint which juxtaposed the expressed V(H)DJ(H) to the C γ 2b region occurred between sequences just 5' to the classical Sμ region and an inverted portion of Sγ2b region followed in turn by a normal Sγ2b sequence. A possible mechanism which could account for the structure is schematically represented in Figure 5.

The 18-8 cell line accumulates large deletions in the S regions during propagation in culture (59). We characterized the nature of one such deletion event that occurred in an 18-8 subclone which had switched to γ2b (59). This deletion event occurred within the Sμ region and one end was located within 100 bp of the switch recombination point. Furthermore, we have found that other A-MuLV transformants which undergo class switch events also rapidly accumulate similar Sμ deletions during propagation in culture. Preliminary evidence also suggests that some lines which have a propensity to switch also accumulate deletions in the S γ 2b region (Alt, unpublished); we have also recently molecularly characterized deletion events within the S γ 2b region in a 300-18 derivitive that had undergone a μ to γ2b switch (Yancopoulos, DePinho, Zimmerman and Alt, submitted). Furthermore, we have found that other

A-MuLV transformants that undergo class switch events also accumulate $S\mu$ deletions during propagation in culture (Kruger and Alt, unpublished). Together, these results suggest a relationship between the deletion phenomenon and the potential for class switching; in a mechanistically similar manner to variable region gene accessibility, class switching may require highly accessible S regions to allow the class switch recombinase to operate. Deletion events prior to class switching may reflect the accessibility of the involved S region to a putative specific switch recombinase; alternatively, they might reflect accessibility to a general recombination system which could also be promoted by the highly repetitive nature of the switch region.

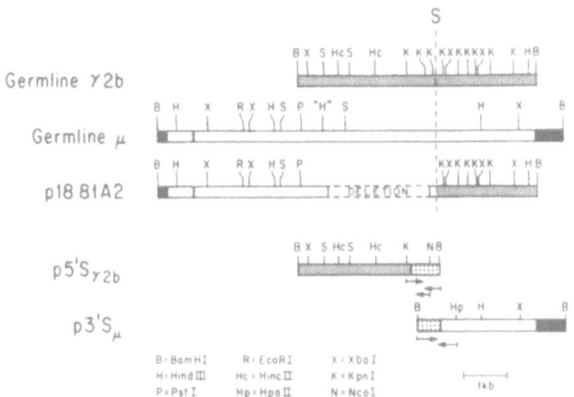

Figure 4. A partial restriction map of the p18.81A2 subclone and comparison with the germline μ and γ2b loci reveals an intra-$S\mu$ deletion as well as a recombination between $S\mu$ and $S\gamma$2b. The intervening region between $S\mu$ and $S\gamma$2b were retained.

Control of Class Switching

Switching to 2b is a high frequency event in pre-B cell lines; nearly all independent transformants thus far characterized which have undergone switch events (five independent lines) have switched to

γ2b (57,58,60). In analogy to our studies of V(H) gene utilization, these switch events also appear to occur spontaneously in the absence of any immunoregulatory or obvious immunoselective influences. Furthermore, we have found significant levels of γ2b-related transcripts in normal fetal liver but no detectable levels of transcripts related to any other C(H) region except Cμ (Yancopoulos, DePinho, Zimmerman, and Alt, submitted). Together, our findings suggest the possibility that preferential switching to γ2b may be a normal part of the pre-B cell differentiation program. Currently, we have no understanding of the potential physiological significance of such a programed class switch event.

Direct switching from μ to γ2b expression could be mediated at the level of the accessibility of the γ2b region to a switch recombinase. This possibility was first suggested by the finding of deletions within these two switch regions in pre-B cell lines which had a propensity to switch from μ to γ2b (see above). The production of so-called sterile transcripts (i.e., C(H) transcripts which do not contain V sequences) from the Cμ gene is a well-known phenomenon (61); recently, these transcripts have been shown to be initiated just upstream from the Sμ region (61). We have also demonstrated that the γ2b constant region is apparently transcribed in cell lines which undergo μ to γ2b switches, leading to the generation of sterile Cγ2b transcripts which lack V regions (42, Yancopoulos, DePinho, Zimmerman and Alt, submitted). By analogy to the sterile transcripts, it seems likely that these transcripts are also initiated upstream from the Cγ2b region. Therefore, both of the involved switch regions appear to be open and transcribed in pre-B cells prior to the switch event, perhaps reflecting a specific increase in the accessibility of these regions to switch-recombination. Others have obtained similar evidence implicating an accessibility mechanism in targeting the switch of a B cell lymphoma line from μ to ε or α (62). These findings raise the possibility that specific regulatory sequences (enhancers?) exist upstream of each C(H) region. Such sequences could be specifically activated to promote a given type of class switch event; for example a Cγ2b-specific enhancer sequence would be activated to promote specific class switch events in pre-B cells. Thus, it seems possible that all directed class switches could be mediated at the level of accessibility, rather than by S region-specific recombinases.

Summary

Our studies of A-MuLV transformed cell lines have provided evidence which suggests that the two distinct recombination systems active during B cell differentiation (variable region recombinase and class switch recombinase) may be regulated by analogous mechanisms which involve modulation of the accessibility of the target gene segments to a "recombinase".

Acknowledgements

We thank Dr. Scott Mellis for critically reading this manuscript. R. A. D. is a recipient of the NIH Physician Scientist Award, NIH AI 00602. This work was supported by an NIH grant ROI CA 40427-01 and Searle Scholars Award to F. W. A., who is a Mallinkrodt Scholar. We thank Rosalie Spata for her expert secretarial assistance.

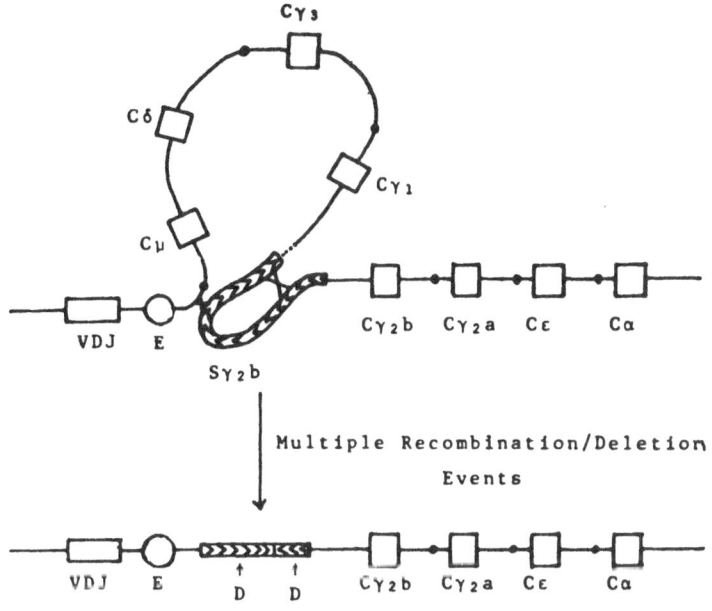

300-18 Pre-B Cell γ₂b Class Switch

(Inversion of Sγ₂b)

Figure 5. A μ to γ2b class switch resulting from a complex recombination/deletion event leading to inversion of Sγ2b. Intra-Sγ2b mini-deletions (D) are indicated.

References

1. Rosenberg, N. and Baltimore, D., A quantitative assay for transformation of bone-marrow cells by A-MuLV, J. Exp. Med. 143:1453 (1976).
2. Yancopoulos, G.D. and Alt, F.W., Regulation of the assembly and expression of variable region genes, Ann. Rev. Immunol., in press.
3. Alt, F.W., Blackwell, T.K., DePinho, R.A., Reth, M.G., and Yancopoulos, G.D., Regulation of genome rearrangement events during lymphocyte differentiation, Immunol. Rev., in press.
4. Givol, D., The antibody combining site, Int. Rev. Biochem. 23:71 (1979).
5. Kabat, E.A., Antibody diversity versus antibody complementarity, Pharm. Rev. 34:23 (1982).
6. Hilschmann, N. and Craig, L.C., Amino acid sequence studies with Bence Jones proteins, Proc. Natl. Acad. Sci. USA 53:1403 (1965).

7. Wu, T.T. and Kabat, E.A., An analysis of the sequences of the variable regions of Bence Jones proteins and myeloma light chains and their implications for antibody complementarity, J. Exp. Med. 132:211 (1970).

8. Tonegawa, S., Somatic generation of antibody diversity, Nature 302:575 (1983).

9. Winkelhake, J.L., Immunoglobulin structure and effector functions, Immunochem. 15:695 (1978).

10. Early, P., Huang, H., Davis, M. Calame, K. and Hood, L., An immunoglobulin heavy chain variable region gene is generated from three segments of DNA: V(H), D and J(H), Cell 19:981 (1980).

11. Sakano, H., Maki, R., Kurosawa, Y., Roeder, W. and Tonegawa, S., Two types of somatic recombination are necessary for the generation of complete immunoglobulin heavy chain genes, Nature 286:676 (1980).

12. Ravetch, J.V., Siebenlist, U., Korsmeyer, S., Waldmann, T. and Leder, P., Structure of the immunoglobulin μ locus: Characterization of embryonic and rearranged J and D genes, Cell 27:583 (1981).

13. Wood, C. and Tonegawa, S., Diversity and joining segments of mouse Ig genes are closely linked and in the same orientation, Proc. Natl. Acad. Sci. USA 80:3030 (1980).

14. Kurosawa, Y. and Tonegawa, S., Organization, structure and assembly of immunoglobulin heavy chain diversity (D) DNA segments, J. Exp. Med. 155:201 (1982).

15. Kemp, D.J., Tyler, B., Bernard, O., Gough, N., Gernodakis, S., Adams, J.M. and Cory, S., Organization of genes and spacers within the mouse immunoglobulin V(H) locus, J. Mol. Appl. Genet. 1:245 (1981).

16. Bothwell, A.L.M., Paskind, M., Reth, M., Imanishi-Kari, T. and Baltimore, D., Molecular basis of a mouse strain-specific anti-hapten response, Cell 24:625 (1981).

17. Loh, D.Y., Bothwell, A.L.M., White-Scharf, M.E., Imanishi-Kari, T. and Baltimore, D., Molecular basis of a mouse strain-specific anti-hapten response, Cell 33:85 (1983).

18. Brodeur, P. and Riblett, R., The immunoglobulin heavy chain variable reigon (Ig-H-V) in the mouse. I. 100 IgH-V genes comprise 7 families of homologous genes, Eur. J.Immunol. 14:922 (1984).

19. Brodeur, P., Thompson, M.A. and Riblett, R., The content and organization of mouse IgH-V families. In Regulation of the Immune System, UCLA Symposia on Molecular and Cellular Biology, New Series, Vol. 18. (New York: Alan R. Liss), pp. 445 (1984).

20. Cory, S., Tyler, B.M. and Adams, J.M., Sets of Ig V(K) genes homologous to ten cloned V(K) genes, J. Mol. App. Genet. 1:103 (1981).

21. Selsing, E. and Storb, U., Mapping of immunoglobulin variable region genes: relationship to the "deletion" model of immunoglobulin gene rearrangement, Nucl. Acids Res. 9:5725 (1981).

22. Sakano, H., Juppi, K., Heinrich, G., and Tonegawa, S., Sequences at the somatic recombination sites of immunoglobulin light chain genes, Nature 280:288 (1979).

23. Max, E.E., Seidman, J.G. and Leder, P., Sequences of five potential recombination sites encoded close to an immunoglobulin region gene, Proc. Natl. Acad. Sci. USA 76:3450 (1979).

24. Einsen, H.N. and Reilly, E.B., Lambda chains and genes in inbred mice, Ann. Rev. Immunol. 3:337 (1985).

25. Alt, F.W., Yancopoulos, G.D., Blackwell, T.K., Wood, C., Thomas, E., Boss, M., Coffman, R., Rosenberg, N., Tonegawa, S. and Baltimore, D., Ordered rearrangement of immunoglobulin heavy chain variable region segments, EMBO J. 3:1209 (1984).

26. Alt, F.W. and Baltimore, D., Joining of immunoglobulin H chain gene segments, implications from a chromosome with evidence of three D-J(H) fusions, Proc. Natl. Acad. Sci. USA 79:4118 (1982).

27. Desiderio, S.V., Yancopoulos, G.D., Paskind, M., Thomas, E., Boss, M.A., Landau, N., Alt, F.W. and Baltimore, D., Insertion of N regions into heavy chain genes is correlated with expression of terminal deoxytransferase in B cells, Nature 311:752 (1984).

28. Blackwell, T.K. and Alt, F.W., Site-specific recombination between Ig D and J(H) segments that were introduced into the genome of a murine pre-B cell line, Cell 37:105, 1984.

29. Yancopoulos, G.D., Blackwell, T.K., Suh, H.-Y., Hood, L. and Alt, F.W., Joining between introduced but not endogenous T-cell receptor variable region gene segments in pre-B cells: Evidence that B and T cells use a common recombinase. Cell, in press.

30. Yancopoulos, G.D. and Alt, F.W., Developmently controlled and tissue-specific expression of unrearranged V(H) gene segments, Cell 40:271 (1985).

31. Banerji, J., Olson, L. and Schaffner, W., A lymphocyte-specific cellular enhancer is located downstream of the joining region in immunoglobulin heavy chain genes, Cell 33:729 (1983).

32. Gilles, S.D., Morrison, S.L., Oi, V.T. and Tonegawa, S., A tissue-specific transcription enhancer element is located in the major intron of a rearranged immunoglobulin heavy chain gene, Cell 33:717 (1983).

33. Neuberger, M.S., Expression and regulation of immunoglobulin heavy chain gene transfected into lymphoid cells, EMBO J. 2:1373 (1983).

34. Mason, J.O., Williams, G.T. and Neuberger, M.S., Transcription cell type specificity is conferred by an Ig V(H) gene promoter that includes a functional consensus sequence, Cell 40:479 (1985).

35. Grosschedl, R. and Baltimore, D., Cell-type specificity of immunoglobulin gene expression is regulated by at least three DNA sequence elements, Cell 41:885 (1985).

36. Reth, M.G. and Alt, F.W., Novel immunoglobulin heavy chains are produced from DJ(H) gene segment rearrangement in lymphoid cells, Nature 312:418 (1984).

37. Pernis, B.G., Chiappino, G., Kelus, A.S. and Gell, P.G.H., Cellular localization of immunoglobulins with different allotype specificities in rabbit lymphoid tissues J. Exp. Med. 122:853 (1965).

38. Cebra, J., Colberg, J.E. and Dry, S., Rabbit lymphoid cells differentiated with respect to $\propto$-, γ- and μ-heavy polypeptide chains and to allotypic markers AA1 and AA2, J. Exp. Med. 123:547 (1966).

39. Alt, F.W., Exclusive immunoglobulin genes, Nature 312:502 (1984.

40. Alt, F.W., Enea, V., Bothwell, A.L.M. and Baltimore, D., Activity of multiple light chain genes in murine myeloma lines expressing a single, functional light chain, Cell 21:1 (1980).

41. Kuehl, W.M., Kaplan, B.A., Scharff, M.D., Nau, M., Honjo, T., Leder, P., Characterization of light chain and light constant region fragments mRNAs in MPC11 mouse myeloma cells and variants, Cell 5:139 (1975).

42. Alt, F.W., Rosenberg, N., Enea, V., Siden, E. and Baltimore, D., Multiple immunoglobulin heavy-chain gene transcripts in Abelson murine leukemia virus-transformed lymphoid cell lines, Mol. Cell. Biol. 2:386 (1982).

43. Reth, M.G., Amirati, P., Jackson, S. and Alt, F.W., Regulated progression of a cultured Pre-B cell line to the B cell stage, Nature 317:353 (1985).

44. Alt, F.W., Rosenberg, N., Lewis, S., Thomas, E. and Baltimore, D., Organization and reorganization of immunoglobulin genes in an A-MuLV transformed cells: rearrangement of heavy but not light chain genes. Cell 27:381 (1981).

45. Hood, L., Kronenberg, M. and Hunkapiller, T., T cell antigen receptor and the immunoglobulin supergene family, Cell 40:225 (1985).

46. Blackwell, T.K., Yancopoulos, G.D. and Alt, F.W., Joining of immunoglobulin heavy chain variable region segments in vivo and within a recombination substrate. In Molecular Biology of Development, UCLA Symposia on Molecular and Cellular Biology, New Series, Vol. 19.
(New York: Alan R. Liss), pp. 537 (1984).

47. Yancopoulos, G.D., Desiderio, S.V., Paskind, M., Kearney, J.F. and Baltimore, D., Preferential utilization of the most J(H) proximal V(H) gene segments in pre-B cell lines, Nature 311:727 (1984).

48. Perlmutter, R.M., Kearney, J.F., Chang, S.P. and Hood, L.E., Developmentally controlled expression of Ig V(H) genes, Science 227:1597 (1984).

49. Manheimer-Lory, A., Monestier, M., Bellon, B., Alt, F.W., Bona, C., Anti-immunoglobulin antibodies VII. V(H) genes and idiotypy of rheumatoid factor, PNAS, in press.

50. Shimizu, A. and Honjo, T., Ig class switching, Cell 36:801 (1984).

51. Marcu, K. and Cooper, M., New views of the immunoglobulin heavy-chain switch, Nature 298:243 (1982).

52. Alt, F.W., Bothwell, A.L.M., Knapp, M., Siden, E., Mather, E., Koshlan, M. and Baltimore, D., Synthesis of secreted and membrane-bound immunoglobulin mu heavy chains is directed by mRNAs that differ at their 3'-ends, Cell 20:293 (1980).

53. Pernis, B., Farni, L. and Luzzati, A.L., Synthesis of multiple immunoglobulin classes by single lymphocytes. Cold Spring Harbor Symp. Quant. Biol. 41:175 (1976).

54. Shimizu, A., Takahashi, N., Yaiota, Y. and Honjo, T., Organization of the constant-region gene family of the mouse immunoglobulin heavy chain, Cell 24:499 (1982).

55. Marcu, K.B., Lang, R.B., Stanton, L.W. and Harris, L.J., A model for the molecular requirements of immunoglobulin heavy chain class switching, Nature 298:87 (1982).

56. Kataoka, T., Kawakami, T., Takahashi, N. and Honjo, T., Rearrangement of immunoglobulin 1 chain gene and mechanism for heavy-chain class switch, Proc. Natl. Acad. Sci. USA 77:919 (1980).

57. Akira, S., Sugiyama, H., Yoshida, N., Kikutani, H., Yamamura, Y. and Kishimoto, T., Isotype switching in murine pre-B cell lines, Cell 34:545 (1983).

58. DePinho, R.A., Kruger, K., Andrews, N., Lutzker, S., Baltimore, D., and Alt, F.W., Molecular basis of heavy-chain class switching and switch region deletion in an Abelson transformed cell line Mol. Cell Biol. 4:2905 (1984).

59. Alt, F.W., Rosenberg, N., Casanova, R.J., Thomas, E. and Baltimore, D., Immunoglobulin heavy chain class switching and inducible expression in an A-MuLV transformed cell line Nature 296:325 (1982).

60. Burrows, R.D., Beck-Engeser, G.B. and Walb, M.R., Immunoglobulin heavy-chain class switching in a pre-B cell line is accompanied by DNA rearrangement, Nature 306:243 (1983).
61. Lennon, G.G., and Perry, R.P., Cμ-containing transcripts initiate heterogeneously within the IgH enhancer region and contain a novel 5' nontranslatable exon, Nature, in press.
62. Stravnezer, J., Abbot, J. and Sirlin, S., Ig H-chain switching in cultured 129 murine B lymphoma cells, committment to an IgA or IgE switch, Curr. Top. Microbiol. Immunol. 113:109 (1984).

ACQUIRED ALTERATIONS AROUND THE Cε GENE IN

CULTURED MOUSE MYELOMA CELLS

Charles-Félix Calvo and Barbara K. Birshtein
Albert Einstein College of Medicine
1300 Morris Park Avenue, Bronx, N.Y.   10461

Our laboratory has focused on the study of various aspects of DNA rearrangements that affect the organization and expression of immunoglobulin heavy chain genes. Using a mouse myeloma cell line, MPC11 (IgG2b,k), we have isolated variants making various types of altered heavy chains, including class switch variants producing γ2a heavy chains, hybrid (γ2b–γ2a) heavy chain producers, and short γ2b heavy chain producers.

Various cultured cell systems have been described in which class switch rearrangements have been detected. In certain of these, DNA rearrangement events have been observed prior to the final switch rearrangement event that leads to new heavy chain gene expression (1). Similarly , in the parental MPC11 prior to variant isolation, there are differences downstream of the expressed γ2b gene when compared to the BALB/c germline constellation. In particular, there is an ∿3.5 kb deletion between the γ2b and γ2a genes that includes a portion of Sγ2a sequences (2) as well as a tandem duplication of the γ2a gene (3). We have considered that these rearrangements in the immunoglobulin cluster in MPC11 have contributed to further productive switch rearrangement.

More recently, we have observed that in at least 60% of the variants we have isolated, including those that make short γ2b heavy chains, there is accompanying copy number variation affecting the γ2a genes on the expressed chromosome, resulting from unequal sister chromatid exchange (3). We thus have examined the region downstream of the γ2a gene for any additional changes that have been acquired. In particular, we have analyzed the region around the Cε gene. Using a probe for the Cε gene, we have carried out genomic Southern analysis on Bam HI restricted DNA from sixteen variants. Fourteen of the sixteen variants had only the germline Cε band of 4.8 kb. Two of the variants, 9.9.1.6.7 and F5.5, showed new bands (Fig. 1). 9.9.1.6.7 makes a heavy chain having a complete γ2a constant region (3) while F5.5 makes a short γ2b heavy chain of 50,000 MW and shows two copies of the J$_H$ region (MPC11 has only one) (G. Gilmore, personal communication).

Additional digests with KpnI and XbaI were carried out. 9.9.1.6.7 showed only the germline Cε band with these enzymes, implying that the

Bam HI alteration reflected a point mutation. In contrast, F5.5 showed a new band (in addition to the germline band) with each of the further digests. Further analysis has localized the region of the new map in F5.5 to 3' of the C ε gene. However, the regions even further 3', flanking the Cα gene, show only the MPC11 bands. Further mapping studies are underway to analyze whether the new C ε 3' map reflects inserted or duplicated DNA or a translocation event.

Two approaches to analyze whether alterations around Cε will bias additional isotype switching in these variant cell lines are designed. The first of these will compare the frequency of switches in culture to IgE production in F5.5 and in a variant cell line that has no such alterations. Another approach will utilize the known enhancement of IgE production in mice infected with Nippostrongylus brasiliensis (4) to provide a possible in vivo "trigger" for IgE production in mice bearing tumors representing the two Cε genotypes.

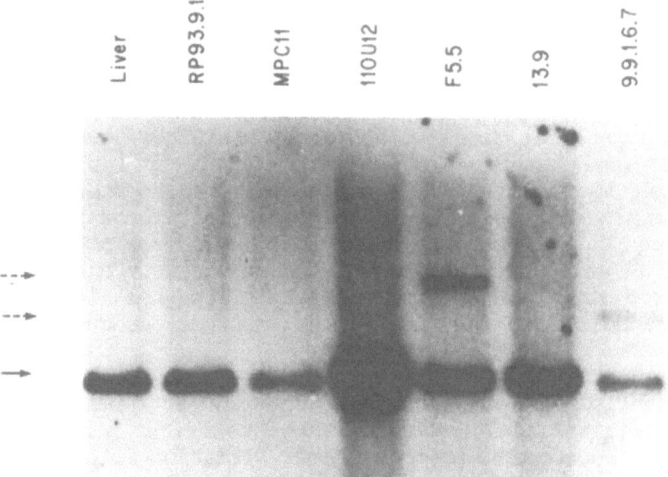

BamHI/cε probe

Fig. 1. Genomic Southern analysis of MPC11 variants showing Cε bands. A BamHI digest was probed with PCε (3.8 kb BamHI-EcoRI fragment isolated from ε-12 insert (5)).
Liver (BALB/c liver DNA); RP93.9.3 (unrelated hybridoma making IgG2a); MPC11 (IgG2b); 110U12 (k only); F55 (short γ2b,k); 13.9 (short γ2b,k); 9.9.1.6.7 (IgG2a).

References

1. Hurwitz, J.L., and J.J. Cebra, 1982. Rearrangements between the immunoglobulin heavy chain gene J_H and C^μ regions accompany normal B lymphocyte differentiation in vitro. Nature (London) 299:742-744.

2. Eckhardt, L.A., and B.K. Birshtein, 1985. Independent immunoglobulin class-switch events occurring in a single myeloma cell line. Molec. Cell. Biol. 5:856-868.

3. Tilley, S.A., and B.K. Birshtein, 1985. Unequal sister chromatid exchange. A mechanism affecting Ig gene arrangement and expression. J. Exp. Med. 162:675-694.

4. Kojima, S., and Z. Ovary, 1975. Effect of Nippostrongylus brasiliensis infection on the anti-hapten IgE response in the mouse. Mechanism of potentiation of the IgE antibody response to a heterologous hapten-carrier conjugate. Cell Immunol. 17:383.

5. Shimizu, A., N. Takahashi, Y. Yamawaki-Kataoka, Y. Nishida, T. Kataoka and T. Honjo, 1981. Ordering of mouse immunoglobulin heavy chain genes by molecular cloning. Nature 289:149-153.

DIFFERENTIATION IN THE I.29 B CELL LYMPHOMA: PRECOMMITMENT TO IgA OR IgE SWITCH IN INDIVIDUAL IgM+ CLONES

R. Sitia, C. Alberini, R. Biassoni, S. DeAmbrosis, and D. Vismara

Laboratory of Molecular Biology,
Istituto Nazionale per la Ricerca sul Cancro
V. le Benedetto XV n.10,
16132 Genova, Italia

INTRODUCTION

B lymphocytes undergo during their development several sequential rearrangements at the immunoglobulin (Ig) loci (1). The first event is a $D-J_H$ recombination that usually takes place at both the heavy chain (IgH) loci, and is then followed by rearrangement of one of the numerous V_H genes to the DJ_H complex (2). When a functional VDJ rearrangement has taken place, synthesis of μ polypeptide ensues and rearragments of the light chain loci begin. A K-λ hyerarchy has been demonstrated in both human and mouse B cells (3,4). A successful heavy chain gene recombination stops any further rearrangement at the IgH locus (5,6), and possibly stimulates IgL recombination. Similarly the creation of a productive light chain gene blocks further recombinations at the IgL loci (5). Subsequently IgM molecules are synthetized and expressed on the cell surface by lymphocytes (7). The latter may coexpress IgD through alternative splicing of the VDJ - $C\mu$ -C δ transcription unit (8). IgM+ B cell may undergo a second recombinatorial event at the IgH locus, termed isotype switching. This event generally implies the deletion of C_μ , C$_\delta$ and all the C_H genes which are located 5' to the one C_H gene that will be expressed (9,10). Although the sequences mediating switch recombination have been isolated and determined (9-11), rather little is known about the mechanisms regulating isotype switching. The question of how an IgM+B lymphocyte decides which C_H gene to recombine is still open. It is well known that certain antigens or routes of immunization result in vivo in the predominance of a given isotype (12). On the other hand several in vitro experiments have shown that more than one isotype can be produced by the progeny of a single B cell (13-15). Isotype switching is not the only differentiative option of an IgM+ B cell. The latter may infact respond to antigen or mitogen by differentiating into IgM secreting plasmacells (7). It is generally accepted that plasmacell may undergo a very limited number of divisions.

To gain information as to the regulatory aspects of B cell differentiation, it would be important to have homogeneous cells lines capable of controlled differentiation in vitro. Here we report characterization of the one such cell line, the murine B cell lymphoma I.29 (16-18). IgM bearing I.29 clones were adapted in vitro and found inducible by LPS to undergo both μ plasmacell differentiation and isotipe switching. Switch recombinations were limited to Cα and C-ϵ genes indicating a precommitment in the differentiative potentials of individual I.29 clones. The molecular basis of this precommitment will be briefly discussed.

MATERIALS AND METHODS

Cell lines, tissue culture and mitogen stimulation. The origin and properties of I.29 μ $^+$ cells (a gift of Dr. Janet Stavnezer, University of Massachusetts, Worcester MA) were described previously [18]. Cells were cultured in RPMI 1640 supplemented with 10% foetal calf serum (FCS), penicillin (100 U/ml) streptomycin (100 ug/ml), non essential aminoacids (1mM), glutamine (2mM), sodium pyruvate (1mM) and, 2-β mercaptoethanol (5.10^{-5}M), and cloned by limiting dilution in the presence of C57Bl/6 peritoneal macrophages (10^4 per well of a 96 plate) as feeder layers. Individual clones were then expanded in the absence of feeder layer cells. For inducing plasmacell differentiation, cell were cultured at 2.10^5/ml in the presence or absence of 5 μg/ml of lipopolysaccharide (LPS) for 3 to 6 days. For inducing isotype switching cells were washed after 4 days of LPS stimulation, and cultured in the absence of LPS until tested by immunofluorescence [19] with class specific antibodies [18-19]. Monoclonal rat anti-mouse ε antibody was a gift from Dr. Eshbar (Rehovot, Israel).

RNA extraction and Northern Blotting. Total cellular RNA was purified by hot phenol (65°C)-chloroform extractions, treatment with RNase-free DNase and ethanol precipitation. Aliquots of 20 μg, measured by spectrophotometry and ethydium bromide staining of test minigels were electrophoresed in 1.2% agarose gels in 10 mM phosphate buffer, pH 7.0 and blotted to either nitrocellulose or diphenylthioether paper [20]. Filter were hybridized with Cμ [21], C-α [21], C ε [22] C$_\gamma$ 1[23] C$^\gamma$2a [24] or I.29-V$_H$ (Klein and Stavnezer, in preparation) specific probes as described previously [20].

RESULTS AND DISCUSSION

LPS may induce IgM secretion and isotype switching in individual I.29 μ^+clones. It was previously shown that membrane IgM bearing cells from the I.29 tumor (I.29 μ^+) could be induced by LPS to differentiate toward IgM secretion (19, Sitia et al., submitted) or switch to the production of IgA, IgE and, less frequently, IgG$_{2a}$ [18].

To determine whether distinct precursors existed within I.29 μ^+ cells for the various differentiative potentials individual μ $^+$ clones obtained by limiting dilution were cultured with or without LPS and analyzed for the presence of μ secreting cells as well as of IgA, or IgE producing cells. All the clones tested (25/25) could differentiate toward IgM secretion, as indicated by the appearance of cells (70%) intensely staining in the cytoplasm with fluorescent anti μ. Approximately 70% of the clones (18/25) could also switch in vitro to the production of IgA and/or IgE. No IgG producing cells were detected in two experiments.

TABLE I.   **Isotype switching and differentiation toward IgM secretion in individual surface IgM$^+$ I.29 clones.**

CLONE	IgM SECRETION[a]	ISOTYPE SWITCHING[b]		
		IgG	IgE	IgA
9	+++	-	-	-
15	+++	-	-	-
16	+++	-	+	±
22	+++	-	+	++

a)   To induce IgM secretion cells were cultured for six days in the presence of LPS and analyzed by cytoplasmic immunofluorescence with anti μ antibodies.

b)   To induce isotype switching cells were cultured for four days with LPS, washed, and cultured in the absence of LPS until tested with class specific fluorescent antibodies.

Table I reports results of four experiments performed on clones 9, 15, 16 and 22. These clones were chosen because the first two never switched in our hands while clones 16 and 22 did; furtheremore clone 16 switched preferentially to IgE, while clone 22 switched preferentially to IgA in response to LPS. These data demonstrate that I.29 μ^+ cells can undertake two mutually exclusive differentiative pathways (plasmacell differentiation and isotype switching) when activated by the same mitogen. As our experiments were performed on cloned B cells, we can exclude any influence of T or accessory cells. It is possible however that this crucial step of B cell differentiation is subjected in vivo to a sophisticated regulation. Several T cell produced factors have been shown active in inducing terminal differentiation toward Ig secretion in both normal and neoplastic B cells or shifting the balance of isotype switching in favor of a given isotype (see 25 for review). A well documented example of the latter class is BCDF γ, which increases the frequency of switches to IgG_1 in LPS activated normal B cells (26). The availability of cloned cells lines capable of differentiating to IgM secretion or switching in vitro will provide an optimal model for dissecting the factors regulating these differentiative events and understanding their mechanism of action.

Individual μ^+ clones are precommitted to switch to a given isotype. Of great interest was the observation of preferential switching to ε or α in clones 16 and 22, respectively. The experiments reported in Table I gave us however rather little information as to the number of actual switch events, which can be overexstimated if the cell cycle of the variant cell is faster than that of the parent line, or underextimated if the reverse were true. To determine more precisely the frequency of switching, cell from clones 15,16 22 and the subclone 22C4 were cultured with or without LPS, washed and then cloned by limiting dilution in the presence of peritoneal C57 Bl/6 cells as feeder layers. Individual clones were then expanded and tested by immunofluorescence. As shown in Table II, the frequency of switching to IgA was quite high in clone 22.

TABLE II. **Precommittents in isotype switching in I.29 u^+ subclones limiting dilution assay[a].**

			Number of clones with Ig phenotype [b]						
EXPT.	CLONE	LPS	TOT	μ	$\mu\alpha$	α	$\mu\varepsilon$	ε	$\alpha\varepsilon$
1	22	-	10	4	0	0	3	3	0
		+	34	18	8	8	0	0	0
2	15	-	12	12	0	0	0	0	0
		+	23	23	0	0	0	0	0
	16	-	12	12	0	0	0	0	0
		+	12	10	0	0	2	0	0
	22C 4	-	12	5	2	0	2	3	0
		+	26	19	7	0	0	0	0

a) Cell were cloned by limiting dilution in the presence of C57Bl/6 peritoneal cells as feeder layer after four days of LPS stimulation (expt.2) and after four days of LPS stimulation followed by four days of culture (expt.1)

b) Determined by cytoplasmic immunofluorescence with class specific antibodies.

In the first experiment, clone 22 cells were cultured 4 days with or without LPS, cultured 4 more days LPS and cloned. At time of cloning, LPS stimulated cells were 16% IgA+ 84% IgM+, while control cells were > 98% IgM < 0.1%IgA+. Sixteen out of 34 clones from LPS stimulated cells were IgA+. Interestingly, 8 of them were composed of variable proportions of IgM+ and IgA+ cells. This implies that switch occurred in these wells after the cloning procedure. In experiment two, cells from clones 15,16 and 22 C4 (the latter being a subclone of 22) were cloned after 4 days of culture in the presence or absence of LPS. Consistent with the data reported in Table I no α or ε swhitches were detected in clone 15, whereas two out 12 wells from LPS-treated clone 16 contained variable proportions of μ^+ and ε cells. Seven/26 wells from LPS-stimulated clone 22C4 contained both μ^+ and α^+ cells. No α only clones were detected in this experiment, suggesting that switching takes place at or after day 4 of LPS stimulation.

Unexpectedly, in clone 22 and its subclone 22C4, a high proportion of wells from cells cloned without prior LPS stimulation were ε^+. The possibility exists that cloning in the presence of allogeneic peritoneal cells stimulated switching to IgE.

The frequency of α switching in clones 22 and its subclone 22C4, as measured by both bulk culture or limiting dilution experiments, was quite similar. Furthermore, the precommitment to switch to a given isotype in response to LPS was maintained in culture for over three months. We also tested two μ^+ subclones isolated from LPS stimulated clone 22: quite unexpectedly they switched to IgA in response to LPS with a frequency similar to the parental clone.

The observation that I.29 μ^+ cells appear "precommitted" in their switching potentials, is in contrast with studies by others who showed that the progeny of single IgM+ normal B lymphocytes can yield antibodies of all classes (13-15). It is possible that switch precommitment in related to the neoplastic nature of I.29 cells. Alternatively, switch precommitment might be a characteristic of a subset or of a particular stage of B cell differentiation (memory cells ?) of which I.29 is the neoplastic representation.

IgE and IgA producing switch variants may further differentiate in response to LPS by increasing the rate of Ig secretion. To verify whether the IgE and IgA switch variants isolated from clones 16 and 22 retained the capability of responding to LPS, the cells from clones 16C5 and 16E10 (ε^+), 22 A11 and 22A17 (α^+) were cultured in the presence of the mitogen and tested by immunofluorescence. The number of cells intensely staining in the cytoplasm increased in all clones (data not shown. See also refs. 19,20,27). No IgA switching could be induced in any of the ε^+ clones.

The subclones precommitted to switch to IgA transcribe the germ line C gene(s). It is generally accepted that expressed genes are hypomethylated and reside in open chromatin domains (28). Recently it has been shown that the germ line V_H genes are transcribed at high rate in pre B cells that are actively rearranging V_H to DJ_H but not in further stages of B cell differentiation (29), implying a correlation between transcriptional competence and availability to recombinatorial events. Previous studies by Stavnezer et al. showed that the $C\alpha$ and, to a lesser extent, the $C\varepsilon$ and $C\gamma_{2a}$ genes are hypomethylated in I.29 μ^+ cells (30-31). Interestingly the IgE+ switch variants had methylated α genes: this is probably the reason why no α switching could be induced in IgE+ I.29 cells.

To determine whether a correlation existed between the chromatin structure of C_H genes and switch precommitment in I.29 subclones, we analyzed C_H gene transcription in clones 22 C3, 22C4, 15 E9 and 16a (all IgM producers), 22 A3 (IgA+), 16C5 (IgE+) and in cells from the K46 B lymphoma (a gift of Dr. Leanderson, Basel Institut for Immunology). As shown in figure 1, panel A, the two IgM+ subclones that showed frequent switching to IgA (22C3 and 22C4 lanes 6 and 7, respectively) contained readily detectable transcripts of these C α gene(s). The size of the α transcripts in clones 22C3 and 22C4 was smaller than that of the mature αmRNAs present in the IgA+ clone 22A3 (lane 1). Furthermore they did not contain VDJ sequences, as shown by hybridization with a VH specific probe (panel C, lanes 6 and 7), suggesting that they represented sterile transcript of either one of the two unrearranged C α genes present in I.29 μ^+ cells. Although we do not have means to assess it, we would predict that these transcripts are encoded by the C α gene located on the expressed chromosome.

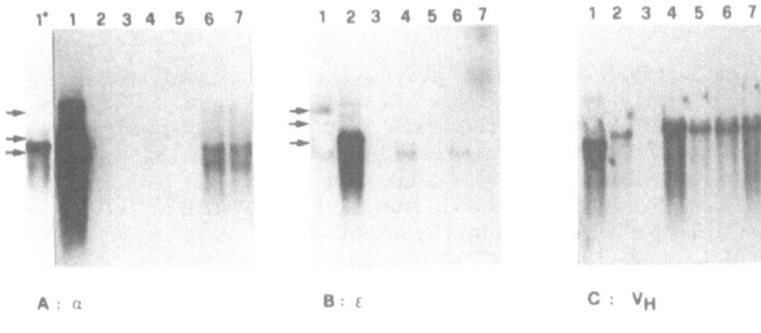

FIG. 1

Aliquots of 20 µg of total cellular RNA were blotted to nitrocellulose filters, hybridized with C_α (panel A), C_ϵ (panel B) and V_H (panel C) probes washed and exposed to Kodak XAR 5 film. **Lane 1:** 22A3 (IgA[+]). **Lane 2:** 16C5 (IgE[+]). **Lane 3:** K46. **Lane 4:** 15E9. **Lane 5:** 16a **Lane 6:** 22C3. **Lane 7:** 22C4. Lane 1* is an underexposure (6 hrs) of lane 6 (24 hrs). Arrows on the left of mRNAs, respectively panels A and B represent mature α (3.1, 2.1 and 1.7 kb and ϵ (3.5, 2.5 and 1.9 kb) respectively.

The possibility that these RNA species arise from processing of a very long transcript initiating from the canonical VDJ promoter (32) cannot be at present formally excluded, but we consider it unlikely in view of the fact that different subclones accumulate different "sterile" transcript, the presence of which correlates well, in the case of α, with the switching properties of the subclones themselves. On the other hand no apparent correlation between switching and the presence of ϵ sterile transcripts was detected (panel B). For instance, IgA[+] 22A3 cells (lane I) contained a 1.5 k.b ϵ RNA specie. This molecules must be encoded by the excluded chromosome, as IgA producing cell have deleted the gene from the expressed chromosome (17-18). Sterile ϵ transcript of similar size were detected also in clone 15E9 (which did not switch in vitro, lane 4) and clone 22C3 (which yielded some IgE producing subclones in the absence of LPS stimulation, lane 6) but not in clone 16 a (a subclone of clone 16.) or in clone 22C4 (whose functional properties were similar to clone 22C3). Hybridization with a V_H specific probe (parel C) confirmed that 15E9, 16a 22C3 and 22C4 all produced mRNAs for both $_\mu$ m and $_\mu$ s.

The switch precommittent in I.29 cells might be also explained by the differential expression of class specific switch recombinases. If this were the case we would expect to detect recombinations involving the same S region on both the expressed and the excluded chromosomes. By contrast, previous studies by Stavnezer et al. have shown either unrearranged C_H genes or recombination of the S $_\gamma$ 3 region on the excluded chromosome of I.29 switch variants (17,18). The results reported in this paper are in agreement with the recent studies of Stavnezer et al., on the uncloned I.29 μ + cell line (18,30,31) and suggest that the chromatin structure of the C_H locus plays a fundamental role in determining isotype switch specificity.

ACKNOWLEDGEMENTS

We are indubted with J.Stavnezer for helpful discussions, for giving I.29µ + cells and for communicating results prior of publication with Drs. B.Birshtein, S.Tilley, D.Klein, T.Honjo, K.Marcu, Z.Eshbar, P.Kincade for probes and antibodies, with Drs. A.Rubartelli and U. Hämmerling for helpful discussions.We thank Dr. G.Vidali for his support and also Mrs. E. Rosellini for typing the manuscript. This work was supported by Progetto Finalizzato Ingegneria Genetica e Basi Molecolari delle Malattie Ereditarie of CNR - Rome, 84.00912.51. R.B. is a recipient of A.I.R.C. fellowship.

REFERENCES

1) Tonegawa S. (1983). Nature 302: 575.
2) Alt F.W. et al. (1981). Cell 27:381.
3) Korsmeyer S. et al. (1981). Proc. Natl. Acad. Sci. USA.
4) Alt. F.W. et al. (1984) EMBO J. 3: 1209.
5) Rusconi S. and G. Kohler (1985) Nature 314: 330.
6) Ritchie K.A., R.L. Brinster and U. Storb (1984). Nature 312:517.
7) Calvert J.A. et al (198) Semin. Hematol.
8) Knapp M.R. et al. (1982) Proc. Natl. Acad. Sci. USA 79: 2996.
9) Sakano H. et al. (1980). Nature.
10) Davis M.M, S.K. and L.E. Hood (1980). Science 209: 1360
11) Kataoka T., T.Mijata and T.Honjo (1981). Cell 23: 357.
12) Katz D.H. (1980) Immunology.
13) Gearhart P.J., N.H. Sigal and N.R. Klinman (1975). Proc. Natl. Acad. Sci. 72:1707.
14) Mongini P., W.E. Paul and E.S. Metcalf (1983) J.Exp. Med.157:69.
15) Teale J.L. (1983) J.Immunol. 131: 2170
16) Sitia R.,A.Rubartelli and U.Hämmerling (1981). J.Immunol 127:1388
17) Stavnezer J.et al. (1982) Mol. Cell. Biol. 2:1002.
18) Stavnezer J., S.Sirlin and J. Abbott (1985) J.Exp. Med. 161:577.
19) Sitia R.et al. (1985) Eur. J. Immunol. 15:570.
20) Sitia R. et al. (1985) J.Immunol. 135: 2859.
21) Marcu K.B. et al (1980) Cell 22:187.
22) Nishida Y. et al. (1981) Proc. Natl. Acad. Sci. USA. 78:1581.
23) Rogers J., P.Clarke and R.W. Salser (1979). Nucleic Acid Res. 6:3305
24) Tilley S.A. and B.Birhstein (1985). J.Exp. Med. 162:675.
25) Moller J. Ed. (1984) Immunol. Rev. Vol 78.
26) Layton J.E. et al. (1984) J.Exp. Med. 160: 1850.
27) Sitia R. (1985). Mol. Immunol. 22:1289.
28) Weintraub H., A.Larsen and G. Groudine (1981) Cell 24:333.
29) Yancopoulos G.D. and F.W. Alt (1985). Cell. 40: 271
30) Stavnezer J. J.Abbott and Sirlin (1984) Curr. Top. Microbiol. Immunol. 113:109.
31) Stavnezer J. and S.Sirlin (1986) EMBO J. (in press).
32) Yaoita Y. et al. (1982) Nature 297:697.

TRANSCRIPTIONAL AND POST-TRANSCRIPTIONAL CONTROL OF Ig-GENE EXPRESSION

IN MURINE B-CELLS ACTIVATED BY LPS AND ANTI-RECEPTOR ANTIBODIES

Una Chen-Bettecken, Eberhard Wecker and Anneliese Schimpl

Institut für Virologie und Immunbiologie der
Universität Würzburg
Versbacher Strasse 7,
D-8700 Wurzbürg, FRG

Bacterial Lipopdysaccharide (LPS) induces normal resting B-cells to proliferate and to differentiate into immunoglobulin (Ig)-secreting plasma
cells. We have studied the transcriptional control of Ig-gene expression
in this system by an in vitro transcriptional run-on assay (1). Fig. 1
shows the relative RNA transcriptional rates in nuclei isolated from
resting B-cells and from B-cells 1 to 4 days after LPS stimulation. There
are very few if any transcripts demonstrable in resting B-cells with the
probes tested (C_μ, Igh-enhancer, and kappa). Even actin and H-2 probes
do not give strong signals. Upon LPS stimulation, there is a rapid and
strong enhancement of RNA polymerase II activities until day 4, giving
30-50 fold increases. The μ- and kappa-transcripts detectable are about
equal, indicating also a balanced distribution of RNA polymerase II
along both heavy and light chain loci. The 30-50 fold increases of
transcription upon LPS induction account for the strong accumulation of
Ig-mRNAs found in day-4 LPS cultures. The data demonstrate that Ig-gene

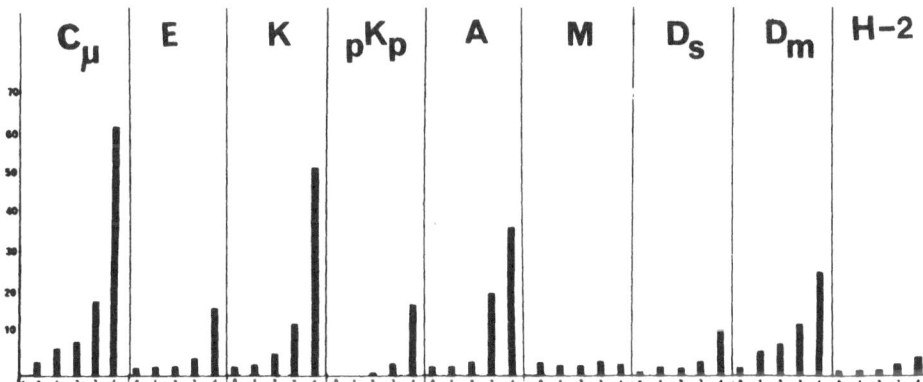

Figure 1. In vitro transcriptional run-on assays of resting and LPS
treated B-cells. The detailed methods, the hybridization probes and the
original references are described in (1,3). C_μu:a μus-cDNA, E:Igh enhancer, K:a Bgl II-Bgl II fragment of kappa locus, pKp:a HindIII-XbaI fragment of kappa intron, A:actin, M:a retrovirus LTR-IS probe, S:28S-rRNA,
Ds:delta-secretory region, Dm:delta-membrane region. H-2:a H-2 classe I
probe.

expression in normal activated B-lymphocytes is regulated primarily at the level of transcription.

We have also studied the activation of B-cells by anti-Ig-receptor antibodies. It is known that anti-μ F(ab')$_2$ fragments (anti-μ) can stimulate B-cells to DNA synthesis but not to Ig-secretion (2). When nuclei from anti-μ treated cells were used for the run-on assay, no significant levels of Ig specific transcripts were detected. This suggests that stimulation by anti-μ fails to provide sufficient signals required for B-cell differentiation or that it inhibits differentiation. The latter possibility was investigated using LPS plus anti-μ (LPS/anti-μ) treatment of B-cells. When anti-μ is added to LPS cultures, the DNA synthesis is not affected, while polyclonal Ig production is inhibited (3), as first described by Melchers and his colleagues (4), and also by Cooper et al. (5).

This phenomenon was further studied at the mRNA level. At the early stage of the culture (day 2), the inhibition by anti-μ seems to be specific for μ-secretory mRNA, since μ-membrane mRNA and kappa-mRNA are not affected (Fig. 2A). At later stages of culture (day 3 to day 4) the pattern of mRNAs becomes more complicated. There is either no μ-specific mRNA left (Fig. 2B), or 1-2 very weak residual bands are still detectable (Fig. 2C). It is interesting to note that in either case the level of kappa-chain mRNA (and J-chain mRNA, not shown) also decrease as a secondary effect, while H-2 mRNA remains unaltered. The remaining μ-mRNA in the mRNA pool can still be translated in vitro into μ-chain polypeptides (Fig. 3). The degree of reduction shown at the protein level is comparable to that shown at the mRNA level.

Subsequently the question was addressed if the specific inhibition exerted by anti-μ treatment occurs at the transcriptional or purely at the posttranscriptional level. As summarized in Table 1, there is no difference in the transcriptional rate between nuclei isolated from day-2 LPS and LPS/anti-μ treated B-cells. Only at a later stage of culture (day 4) there is a slight reduction of the transcriptional rates in the LPS/anti-μ treated cells: A twofold difference with Cμ and Ck probes and a 4.5 fold difference with the Igh-enhancer probe. However, these 2-5 fold differences at the transcriptional level can not account for all the differences detected at the mRNA level (3).

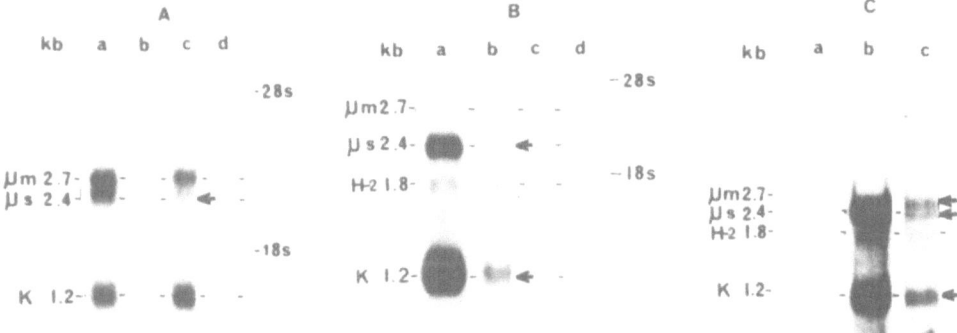

Figure 2: Northern blot analysis of mRNA after 2 days (Fig. 2A) stimulation of small resting B-cells with LPS (lane a), anti-μ (lane b), LPS/anti-μ (lane c), and control culture with medium (lane d). Fig. 2B shows RNA blot analysis of mRNA of B-cells after 4 days culture with LPS (lane a), LPS/anti-μ (lane b), anti-μ (lane c) and control culture (lane d). Fig. 2C shows another blot analysis of mRNA of B-cells after 4 days stimulation with anti-μ (lane a), LPS (lane b), LPS/anti-μ (lane c).

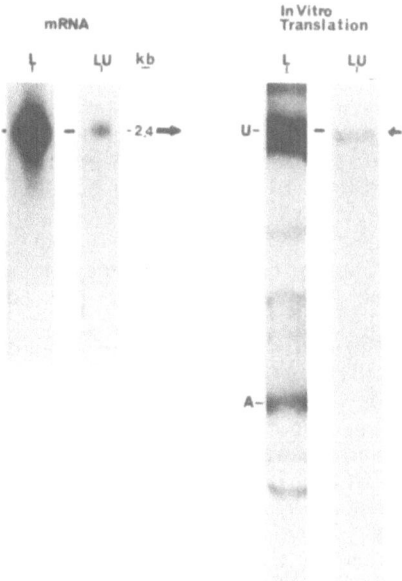

Figure 3. In vitro translation. One µg each of poly-U column selected mRNA from LPS (L) and LPS/anti-µ (LU) B-cells cultured for 4 days were translated in a rabbit reticulocyte lysate translation system (6), then immunoprecipitated with goat anti-mouse IgM antibodies and electrophoresed through a 12.5 % SDS polyacrylamide gel. U:µ-chain, A:actin.

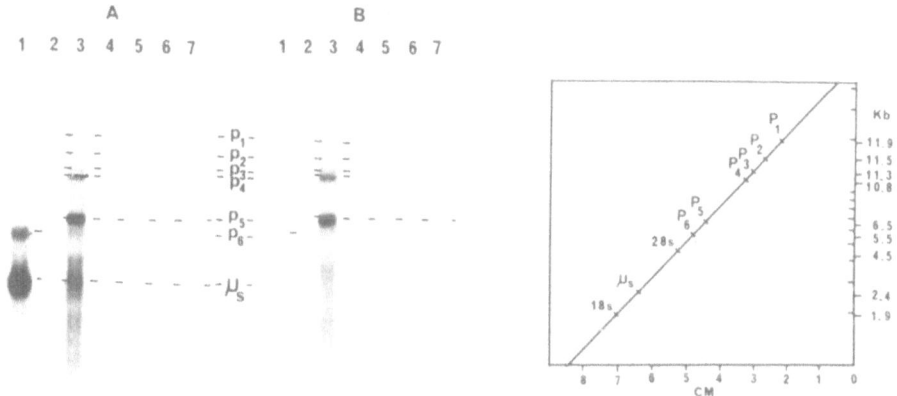

Figure 4. Northern blot analysis of total RNA after 4 days culture of B-cells with: LPS (lane 1), LPS/anti-µ (lane 2), LPS/anti-µ/r-interferon (lane 3), LPS/anti-µ/r-interferon/IL-2 (lane 4), LPS/anti-µ/TRF (lane 5), LPS/anti-µ/BCGF (lane 6), & LPS/anti-kappa F(ab')$_2$ (lane 7). Fig. 4A RNA blot was hybridized with a µ$_s$-cDNA probe, & Fig. 4B blot was hybridized with a µ-intron (Igh-enhancer) probe (2).

Next it was attempted to rescue the LPS activated B-cells from anti-/u suppression by various means. So far only large numbers of T-cells (anti-M locus) can completely revert the negative effect. Lymphokines of different sources were also used to abrogate the suppression (Fig. 4). Among the lymphokines, r-interferon had the most prominent effect. It can stabilize at least five hnRNA precursors ranging from 6.5 Kb to 12 Kb. In addition it leads to the partial reappearance of mature /u-mRNA.

We conclude that the transcription of Ig-genes is greatly enhanced when resting B-cells are activated by LPS. The addition of anti-receptor antibodies to LPS activated B-cells specifically inhibits accumulation of Ig-related mRNAs and thus corresponding protein synthesis. The initial inhibitory effect seems to be specific for the /u-secretory mRNA, then, as a secondary event, kappa-chain and δ-chain mRNAs are also reduced. This anti-differentiation effect is primarily a post-transcriptional event with a minor reduction of Ig transcription. There are several possible post-transcriptional mechanisms which could be responsible for this negative regulation. Differential mRNA stability is one plausible mechanism. However, we favor the hypothesis that anti-/u treatment primarily interferes with the proper /u-RNA processing occurring already in nuclei, either by producing aberrant and unstable primary Ig-transcripts and/or by inhibiting the RNA splicing apparatus.

Table 1. The comparative ratios of in vitro transcriptional rates between nuclei from LPS and LPS plus anti-/u treated cells isolated on day 2, 3 and 4 of cultures.

Ratio LPS / LPS plus anti-/u at locus

Day of isolation	$C_{/u}$	Enhancer	C_k	actin	H-2
day 2	1,00	-	1,07	-	-
day 3	2,18	2,12	1,58	0,99	0,98
day 4	2,02	4,52	2,80	1,10	-

REFERENCES

1. Mather, E.L., Nelson, K.J., Haimovich, J. and Perry, R.P. (1984). Cell 36, 329-338.
2. Parker, D.C. (1980). Immunol. Rev. 52, 115.
3. Chen-Bettecken, U., Wecker, E. and Schimpl, A. (1985). Proc. Natl. Acad. Sci. USA November.
4. Andersson, J., Bullock, W.W. and Melcher, F. (1974). Eur. J. Immunol 4, 715-722.
5. Webb, C.F., Gatings, W.E. and Cooper, M.D. (1983). Eur. J. Immunol. 13, 556-559.
6. Pelhem, H.R.B. and Jackson, R.J. (1976). Eur. J. Biochem. 67, 247-256.

RESTRICTED SET OF V GENES AND EXTENSIVE IDIOTYPE CROSSREACTIVITY OF AUTOANTIBODIES

Marc Monestier, Catherine Painter, Audrey Manheimer-Lory, Blanche Bellon and Constantin Bona

Mount Sinai Medical Center
One Gustave Levy Place
New York, New York 10029

The production of antibodies recognizing self-antigens is one of the hallmarks of autoimmune diseases (1). Despite the obvious role of T cells in the production of autoantibodies and the existence of genetic and environmental factors, the investigation of autoreactive B cell clones may bring important clues in the comprehension of the pathogenesis of autoimmunity. One of the rational approaches is to look for the existence of common features among autoantibodies of various specificities. The existence of similarities could imply that common mechanisms are involved in diseases with different clinical manifestations.

For instance, crossreactive idiotypes (IdX) can be observed on antibodies during a normal immune response (2). This idiotypic crossreactivity seems to be particularly frequent among autoantibodies. Crossreactive idiotypes have been detected among autoantibodies to thyroglobulin (3,4), DNA (5,6,7), Sm (8), acetylcholine receptor (9), and rheumatoid factors (10,11). Such crossreactive idiotypes may function as targets for specific suppression by antiidiotype antibodies (12,13).

The production of autoantibodies may be restricted to a B cell lineage bearing Ly.1 antigen. The Ly.1 antigen was originally described on the T cell membrane and was believed to be a marker of all subsets. However, Ly.1 antigen was later identified on a small percentage of B cells (2%) in normal mice. This percentage is higher (5-10%) in spleen cells of autoimmune NZB mice. The Ly.1 B cells appear early during ontogeny and display an immature phenotype (14). These cells spontaneously secrete IgM in vitro, and in NZB mice, the Ly.1 B cells contain most of anti-DNA, anti-thymocyte, and anti-red blood cell antibody-secreting cells (15). The progenitors of these Ly.1 B cells are different from those of regular B cells and thus, Ly.1 B cells seem to follow a distinct developmental lineage (16).

In this paper we summarized the data obtained in our laboratory concerning the molecular and immunochemical characteristics of autoantibodies with various specificities.

I. AUTOANTIBODIES OF VARIOUS SPECIFICITIES FREQUENTLY EXPRESS CROSS-REACTIVE IDIOTYPES

We have mentioned above the frequency of crossreactive idiotypes among autoantibodies. Until now the presence of one particular IdX had been investigated only within one antigenic specificity. Our studies were aimed to investigate whether or not the IdX borne by autoantibodies of one specificity are shared by other autoantibodies with a different specificity.

We used three different systems in which the idiotypes were characterized with affinity purified anti-Id antibodies. In the first, the IdX were borne by monoclonal mouse rheumatoid factors obtained from various strains: normal BALB/c after activation with lipopolysaccharide, MRL/lpr which develop a lupus-like disease, and 129/Sv which secrete rheumatoid factors without clinical symptoms. The second system used IdX borne by monoclonal mouse anti-Sm antibodies (anti-Sm antibodies are anti-ribonucleoprotein antibodies present in patients with systematic lupus erythematous and in MRL/lpr mice). The last group used IdX present on anti-thyroglobulin antibodies obtained from BALB/c mice injected with thyroglobulin.

Using a competitive radio-immune assay, we investigated the presence of these IdXs among a panel of monoclonal mouse autoantibodies of various specificities: DNA, thyroglobulin, red blood cell, collagen, Sm, membrane receptors, cellular antigens, and self-immunoglobulins. Twenty three of these autoantibodies shared IdXs originally borne by a mouse monoclonal rheumatoid factor. It should be mentioned that in a panel of 16 monoclonal antibodies directed against a foreign antigen (influenza virus), none of them were shown to bear these rheumatoid factor IdX. Eleven autoantibodies were positive in the system defined by anti-Sm antibodies. On the other hand, the third IdX system was shown to be only present among anti-thyroglobulin antibodies of the panel. The expression of these IdX is independent of the specificity, the V_H family used, the mouse strain, and whether the autoantibodies were induced or produced spontaneously.(Table 1)

Obviously, this wide idiotypic crossreactivity of autoantibodies should be significant for the regulation of B cells producing autoantibodies. One may envision several scenarios:
  a) Crossreactive idiotypes could be the target of idiotype specific suppressor T cells and thus play an important role in the tolerance against self antigens.
  b) They could be the target of helper T cells expanded prior to the onset of disease; thus these T cells could collaborate with B cells specific for autoantigens. This can explain the appearance of non-organ specific antibodies which are observed in systemic autoimmune diseases. Indeed, we observed a large idiotype crossreactivity among RF, anti-DNA and anti-Sm antibodies produced by MRL/lpr mice.

Electrophoretic separation of the heavy and light chains of autoantibodies followed by nitrocellulose transfer and probing with polyclonal anti-Id antibodies showed that IdX could be borne by both heavy and light chains.

IdX are often markers of a particular V_H family as was shown for anti-dextran, anti-NP or anti-arsonate antibodies (17). Moreover, IdX have been demonstrated among antibodies of various specificities that were derived from the same V_H family: V_H J558 (18). These data prompted us to investigate the molecular basis of our autoantibodies, namely, the V_H families utilized by them.

34

Table 1. Autoantibodies express crossreactive idiotypes
(the data are expressed as the ratio of IdX
positive antibodies to total antibodies).

	Rheumatoid factor IdX	Anti–Sm IdX	Anti–thyroglobulin IdX
Rheumatoid factors	10/20	8/20	1/20
Anti–thyroglobulin	2/9	0/9	4/9
Anti–DNA	2/5	0/5	0/5
Anti–Sm	2/4	3/4	0/4
Anti–red blood cell	1/3	0/3	0/3
Anti–TSH receptor	1/1	0/1	0/1
Anti–collagen	1/10	0/10	0/10
Anti–acetylcholine receptor	1/1	0/1	0/1
Anti–microfibrils	1/2	ND	0/2
Anti–skin antigens	2/3	ND	0/3
Anti–influenza virus	0/16	ND	ND

II. AUTOANTIBODIES USE A RESTRICTED SET OF V_H GENES

It is established that the V_H genes coding for the variable region of the heavy chains of immunoglobulins can be divided into at least 7 families according to sequence homology (19).

We investigated the V_H families used to encode 43 autoantibodies of our panel. After transfer to nitrocellulose of cytoplasmic lysates or electrophoretically separated RNA, we incubated the filters with a probe for each of the seven V_H families. (Table 2).

We observed that 2 major families encode these autoantibodies: J558 encodes 17 and 7183 which encodes 20 of the 43 antibodies. One antibody uses a V_H gene from the S107 family and five use the QPC57 family. The importance of the J558 family is not surprising since this is the largest family and it comprises about half of the murine V_H germline genes. Of interest is the role of the 7183 family: this family includes less than 10% of the V_H genes but almost half of the autoantibodies tested used $V_H 7183$ genes. $V_H 7183$ is also the most 3' of the V_H families and therefore closest to the D region. It is infrequently utilized by mature cells, but these J_H-proximal V_H gene segments are used preferentially to form $V_H DJ_H$ rearrangements in pre-B cell lines (20). Hybridomas obtained from fetal lines obtained between 16 and 19 days of gestation show also a preferential utilization (10 out of 16) of the $V_H 7183$ gene family (21). All these data suggest that the V_H repertoire of early B-lineage cells is largely restricted to the $V_H 7183$ family. The adult repertoire utilizes all V_H families but our autoantibodies, although obtained from adult mice, show a significant difference in the use of V_H7183. Thus, one may envision that the autoactive B cell clones may possess an immature phenotype and that the self-reactive repertoire is present during ontogeny. During normal development, these self-reactive clones are repressed, but several factors (aging, viral infection, genetic background) may trigger their expansion.

Table 2. V_H gene families expressed in autoantibodies

V_H family	V_H X24	V_H J606	V_H 36-60	V_H J558	V_H S107	V_H QPC52	V_H 7183
Rheumatoid factors				4		2	7
anti-thyroglobulin				4		1	4
anti-DNA				3			2
anti-Sm				3	1		
anti-red blood cells							3
anti-TSH receptor							1
anti-collagen				3		1	1
anti-microfibrils						1	
anti-skin antigens							2
frequency	0/43	0/43	0/43	17/43	1/43	5/43	20/43

The mechanisms that regulate the normal suppression of autoreactive clones are actually unknown but it is possible that the V_H7183 products on the surface of the B cells are targets for this suppression or that their expression is associated with other markers leading to the repression of the clones. For instance, we have mentioned the existence of the Ly.1 marker on B cells which display an immature phenotype and secrete autoantibodies. The V_H7183 gene may be preferentially used by the Ly.1 subset of B cells.

Since we observed such a predominance of V_H7183 usage among autoantibodies, it was of interest to determine the percentage of autoantibodies among a representative sample of antibodies using a gene from the V_H7183 family.

III. ANTIBODIES USING V_H7183 GENE FAMILY ARE MAINLY SELF-REACTIVE ANTIBODIES

After LPS activation in vitro during 48 hours, mouse splenocytes were fused to obtain hybridomas. The hybridomas produced according to this technique are known to randomly use V_H genes (22). Cytoplasmic lysates were prepared, transferred to nitrocellulose and hybridized with a V_H7183 probe. Each V_H7183-positive hybridoma was cloned and the procedure repeated on each clone. The use of a normal length V_H7183 mRNA by each positive clone was confirmed by Northern-blotting technique.

From 3 strains of mice (BALB/c, NZB, and MRL/lpr), we obtained a total of 35 hybridomas using a V_H gene from the 7183 family. The frequency of these 7183 hybridomas was higher in the NZB autoimmune strain (15.8%) than in BALB/c (3.9%).

These 35 hybridomas were expanded and monoclonal antibodies were purified from the supernatant by affinity chromatography. These antibodies were then tested against a variety of self-antigens using several techniques (radio-immunoassay, enzyme-linked immunosorbent assay, immunofluorescence). The specificities tested included nuclear antigens (DNA, Sm), thyroglobulin, immunoglobulins, myelin basic protein, tissue antigens (collagen, glomerular basement membrane, smooth muscle), and cell surface antigens (red blood cell, thymocyte).

Out of 35 antibodies, 5 were rheumatoid factors, one bound to thyroglobulin, 7 to Sm, 4 to DNA, and one to glomerular basal membrane. This binding was specific since it was antigen-inhibitable.

Five other antibodies exhibited various binding specificities for several self-antigens. The binding of these multi-specific antibodies was not antigen inhibitable. All results taken together, about 60% of the V_H7183 antibodies bound to self-epitopes. It should be also noted that 27 of 35 antibodies shared IdX with LPS10-1, a monoclonal rheumatoid factor. Two antibodies of our panel were found to bind to any antigen tested, including BSA. We named these antibodies "sticky" and their physiological role is yet unknown. (Table 3).

All the experiments described above showed the reciprocal correlation between autoantibodies and the V_H7183 family, i.e., a high percentage of autoantibodies use a V_H gene from the 7183 family, and most of the V_H7183 antibodies are self-reactive. The causes of this correlation and the mechanisms leading to the physiological repression of autoreactive B cells remain to be discovered.

These collective results of self-reactive B cells indicate that B cells producing autoantibodies use a restricted set of V_H gene families and that a high percentage of antibodies using a V_H gene derived from the V_H7183 family of germline genes exhibit specificity for self-antigens. Autoantibodies share crossreactive idiotypes which are expressed independently of MHC and Ig gene complexes. These crossreactive idiotopes probably have important regulatory functions as targets either of suppressor T cells involved in the maintenance of self-tolerance or in the activation of autoreactive clones by helper T cells in various circumstances in which self-tolerance is broken.

Our study of the specificity of autoantibodies indicate that they can be classified into three major categories.
a) Monospecific autoantibodies (antigen-inhibitable). These antibodies have high affinity for the self-epitopes, are found in organ-specific autoimmune diseases, and are directly responsible for the tissue injury. Examples of these antibodies are anti-acetylcholine receptor antibodies (myasthenia gravis), anti-red blood cell antibodies (autoimmune hemolytic anemia), anti-thyroid antibodies (thyroiditis).
b) Multispecific antibodies. (non-antigen inhibitable). The binding of these antibodies to self-antigens cannot be explained by a specific paratope-epitope interaction because the paratope of a given antibody cannot accommodate the binding of diverse self-antigens which do not share a common epitome. A more plausible explanation of the multispecificity is that the binding is not via the classical interaction of the antigen and the antibody combining site, but rather that the molecular interaction is via weak bonds (electrostatic or hydrophobic forces) between external surfaces of 3-dimensional structure of variable domain of autoantibodies and the regions exposed on the surface of antigenic molecules. These multispecific autoantibodies could have a housekeeping or scavenger function contributing to the clearance of altered self-macromolecules, aged cells, or even metabolites, as was proposed by Grabar(23). These autoantibodies,

Table 3. Binding properties of monoclonal antibodies produced by hybridomas selected with V_H7183 probe

Origin of antibodies	MRL/lpr	NZB	BALB/c
Nbr of V_H7183^+	5	16	14
Nbr of V_H7183^+ secreting Ig	5	14	14
Autoantibodies antigen-inhibitable	5	6	6
Autoantibodies with multiple specificities	0	2	3
"Sticky" antibodies	0	1	1
Antibodies specific for foreign antigen	0	0	2
Antibodies with un-known specificity	0	5	2

subsequent to a somatic mutational process, could acquire high affinity for a single self-epitome, thereby initiating an autoimmune process.

c) "Sticky" antibodies represent an immunochemical enigma which probably will be solved by precise immunochemical studies.

Further immunochemical studies investigating the binding of autoantibodies to foreign antigens could provide new insight into the mechanisms contributing to the breaking of self-tolerance which is an essential requirement for autoimmunity.

ACKNOWLEDGEMENTS

This study was supported by Grant AG/A1271601 from U.S. Public Health Service. M. Monestier is supported by Fondation pour la Recherche Medicale and the French Government.

REFERENCES

1. Autoimmunity: genetic, immunologic, virologic and clinical aspects N. Talal, ed., Academic Press, New York (1977).
2. Immune Networks, Bona, C., and Kohler, H. ed., Ann. NY Acad. Sci., Vol. 418, (1983).
3. J.P. McCoy, J.H. Michaelson, and P.E. Bigazzi, Anti-idiotypic immunity III. Investigations in human autoimmune thyroiditis. Life Science. 32:109 (1983).
4. M. Zanetti, M. De Baets, and J. Rogers, High degree of idiotypic crossreactivity among murine monoclonal antibodies to thyroglobulins, J. Immunol. 131:2452 (1983).
5. R.S. Schwartz, and D. Stollar, Origins of anti-DNA autoantibodies, J. Clin. Invest. 75:321 (1985).

6. R. Halpern, A. Davidson, A. Lazo, G. Solomon, R. Lahita, and
    B. Diamond, Familial systemic lupus erythematosus, Presence of a
    crossreactive idiotype in healthy family members. J. Clin. Invest.
    76:731 (1985).

7. B.H. Hahn, and F.M. Ebling, A public idiotypic determinant is
    present on spontaneous cationic IgG antibodies to DNA from mice
    of unrelated lupus prone strains, J. Immunol. 133:3015 (1984).

8. D.S. Pisetsky, and E.A. Lerner, Idiotypic analysis of a monoclonal
    anti-Sm antibody, J. Immunol. 129:1489 (1982).

9. D.S. Dwyer, R.J. Bradley, C.K. Urquhart, and J.F. Kearney, Natural-
    ly occurring antiidiotypic antibodies in myasthenia gravis patients,
    Nature (London) 301:611 (1983).

10. D.W. Andrews, and J.D. Capra, Complete amino acid sequence of vari-
    able domains from two monoclonal human anti-gammaglobulins of the
    Wa cross-idiotypic group: suggestion that the J segments are in-
    volved in the structural correlate of the idiotype, Proc. Natl.
    Acad. Sci. USA, 78:3799 (1981).

11. P.P. Chen, S. Fong, D. Normansell, R.A. Houghten, J.G. Karras,
    J.H. Vaughan, and D.A. Carson, Delineation of a crossreactive
    idiotype on human autoantibodies with antibody against a synthetic
    peptide, J. Exp. Med. 159:1502 (1984).

12. E. Neilson, and S.N. Phillips, Suppression of interstitial nephritis
    by autoantiidiotypic immunity, J. Exp. Med. 155:179 (1985).

13. M. Zanetti, and P.E. Bigazzi, Antiidiotypic immunity and autoimmunity
    I. in vitro and in vivo effects of antiidiotypic antibodies to
    spontaneously occurring autoantibodies to rat thyroglobulin.
    Eur. J. Immunol. 11:187 (1981).

14. K. Hayakawa, R.R. Hardy, D.R. Parks, and L.A. Herzenberg, The "Ly-1
    B cell" subpopulation in normal, immunodefective and autoimmune
    mice, J. Exp. Med. 157:202 (1983).

15. K. Hayakawa, R.R. Hardy, M. Honda, L.A. Herzenberg, A.D. Steinberg
    and L.A. Herzenberg, Ly.1 B cells: functionally distinct lympho-
    cytes that secrete IgM autoantibodies, Proc. Natl. Acad. Sci.,
    USA., 81:2494 (1984).

16. K. Hayakawa, R.R. Hardy, L.A. Herzenberg, and L.A. Herzenberg,
    Progenitors for Ly.1 B cells are distinct from progenitors for
    other B cells. J. Exp. Med. 161:1554 (1983).

17. The Biology of Idiotypes, M.I. Green and A. Nisonoff, eds. Plenum
    Press, New York, London p19-59 (1984).

18. C. Victor-Kobrin, T. Manser, T.M. Moran, T. Imanishi-Kari, M. Gefter,
    and C.A. Bona, Shared idiotopes among antibodies encoded by heavy
    chain variable region (V_H) gene members of the J558 V_H family as
    basis for crossreactive regulation of clones with different anti-
    gen specificity, Proc. Natl. Acad. Sci. USA., 82:7696 (1985).

19. P. Brodeur, and R. Riblet, The immunoglobulin heavy chain variable
    (Igh-V) locus in the mouse I. 100 Igh-V genes comprise 7 families
    of homologous genes. Eur. J. Immunol. 14:922 (1984).

20. G.D. Yancopoulos, S.V. Desiderio, M. Paskind, D. Kearny, D. Balti-
    more and F.W. Alt, Preferential utilization of the most J_H
    proximal V_H gene segments in pre B cell lines. Nature (London)
    311:727 (1984).

21. R.M. Perlmutter, J.F. Kearney, S.P. Chang and L.E. Hood, Develop-
    mental controlled expression of immunoglobulin V_H genes. Science,
    227:1597 (1985).

22. R. Dildrop, U. Krawinkel, E. Winter, and K. Rajewski, V_H gene expres-
    sion in murine lipopolysaccharide blasts distributes over the
    nine known V_H gene group and may be random. Eur. J. Immunol.
    15:1154 (1985).

23. P. Grabar, Autoantibodies and the physiological role of immuno-
    globulins. Immunol. Today, 12:337 (1983).

ANTIGEN-SPECIFIC STIMULATION OF HUMAN AUTOREACTIVE B LYMPHOCYTES

T. Logtenberg, A. Kroon, F. H. J. Gmelig-Meyling, and
R. E. Ballieux

Department of Clinical Immunology, University Hospital
Catharijnesingel 101, 3511 GV Utrecht, The Netherlands

INTRODUCTION

The generation of antibody forming cells (AFC) after antigen-specific
stimulation of B lymphocytes involves a number of distinct, separately
controlled steps. These are transition from G0 to G1 of the cell cycle
after binding of antigen to surface immunologlobulin (1) (activation),
several cycles of cell division (proliferation) and finally maturation
into immunoglobulin secreting cells (2). For these steps, most B cell
responses require participation of T cells or T cell-derived lymphokines which may act in a MHC restricted or MHC non-restricted manner
(3,4).

At present it is still unknown if the above steps are also involved
in the differentiation pathway of autoreactive B cells. Although physi-
cally present in healthy individuals, these B cells may have been 'si-
lenced' through previous contact with autoantigen (5). This could result in a quantitatively and/or qualitatively different requirement for
activation, proliferation and differentiation signals. To approach this
question we studied the in vitro response of lymphocytes, from patients
with autoimmune thyroiditis and from healthy individuals, to the
autoantigen thyroglobulin (Tg) and to the xenoantigen ovalbumin (OA).

MATERIALS AND METHODS

Mononuclear cells (MNC) were obtained from blood of patients with
autoimmune thyroiditis and from blood, tonsil and spleen of individuals
without autoimmune disease ('healthy individuals'). T cells were removed by 2 cycles of sheep erythrocyte rosetting and cytotoxic treatment with E7, a pan-T monoclonal antibody, plus complement. Non-T pre-

parations contained <0.1% T cells as judged by OKT11 immunofluorescence. The cells were cultured at 1×10^6 per well in 24-well flat-bottom plates in the presence of free or affigel-coupled antigens or the polyclonal B cell stimulators Pokeweed mitogen (PWM) and Staphylococcus aureus (Sta) (6,7). After 6 days Ig secreting cells were enumerated in a spot Elisa assay as described elsewhere (6,7). Human Tg and OA were coupled with affigel 10 and affigel 15 beads, respectively (6,7). Mitogen-free supernatant of a human T-T hybridoma (Ko 2.9, T. Logtenberg et al., manuscript in preparation) was used as a source of B cell growth and differentiation factors (BGDF). This supernatant gives excellent proliferation and differentiation of Sta preactivated human B cells using standard procedures (8).

RESULTS AND DISCUSSION

The presence of autoreactive B cells in blood, tonsil and spleen was demonstrated by polyclonal stimulation of MNC with a combination of PWM and Sta, assaying both the numbers of IgM and IgG anti-Tg AFC and the numbers of total IgM and IgG forming cells. As shown in Table 1, IgG and IgM anti-Tg AFC were detected in cultures of patients' blood MNC, whereas MNC from healthy individuals, irrespective of their source, invariably produced IgM anti-Tg only.

Table 1. Mitogenic stimulation results in the generation of IgG and IgM anti-Tg and IgM anti-Tg AFC in cultures of MNC from patients and normals respectively. MNC were cultured with a combination of PWM + Sta for 6 days and assayed for numbers of IgM and IgG anti-Tg and total IgM and IgG AFC in the spot Elisa. Results are the mean of duplicate cultures.

| | | Spots/10^6 MNC | | | |
		IgM anti-Tg	IgG anti-Tg	Total IgM	Total IgG
Patients	Exp.				
Peripheral blood	1	245	675	35.000	74.500
	2	200	360	18.400	32.200
Normals					
Peripheral blood	1	55	<1	12.250	16.500
	2	260	<1	39.300	49.000
Tonsil	1	310	<1	24.240	17.200
	2	130	<1	31.600	24.400
Spleen	1	165	<1	18.600	28.600
	2	95	<1	36.800	42.400

We subsequently turned to antigen-specific stimulation. Pilot experiments revealed that stimulation of patients' MNC with <u>soluble</u> Tg did result in the generation of IgG anti-Tg AFC, but only in a minority of the cases tested. We therefore coupled Tg with a solid matrix (i-Tg), as described, and analyzed the response of patients' B cells to the autoantigen thus presented. The results, which are illustrated by the outcome of one representative experiment (Fig. 1), can be summarized as follows. The B cells from 16 out of 19 patients tested could be stimulated with i-Tg to generate anti-Tg AFC. This response could be blocked by the presence of a high dose of soluble Tg (100 µg/ml) during culture. Addition of exogeneous BGDF only resulted in a slight increase in the numbers of IgG anti-Tg AFC in part of the experiments. No IgM α-Tg AFC were generated under these experimental conditions. Growth and differentiation factors present in FCS did not contribute to the stimulation, because the batch of FCS used in these studies does not support the differentiation of Sta preactivated B cells.

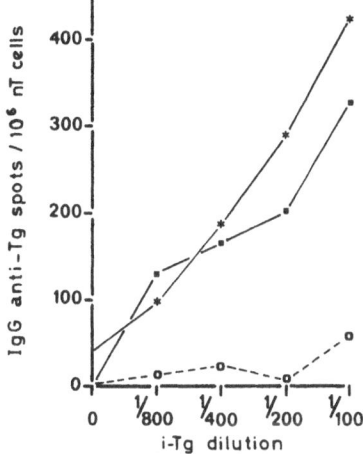

Figure 1. B cells were cultured with serial dilutions of i-Tg with or without 100 µg/ml free-Tg or BGDF (10% T-T hybridoma supernatant). Results are the mean of duplicate cultures.
✱ : + BGDF   ■ : no addition   □ : + 100 µg/ml free-Tg

Next, we analyzed the response of normal human tonsil, spleen and blood B cells both to i-Tg and, as a control antigen, to OA coupled with affigel beads (i-OA). Table 2 depicts the results of a representative experiment with spleen B cells. We found that tonsil and spleen, but not blood B cells (n=10), generate IgM AFC after stimulation both by the autoantigen Tg and by the xenoantigen OA, coupled with affigel

Table 2. Stimulation of spleen B cells with insolubilized Tg or OA and BGDF.

Stimulation		Number of spots/10^6 nT cells		
Antigen	BGDF	IgM anti-Tg	IgM anti-OA	Total IgM
-	-	<1	<1	1.275
-	+	<1	<1	3.780
1/100 i-Tg	+	138	<1	3.940
1/200 i-Tg	+	110	<1	3.620
1/100 i-OA	+	<1	196	4.260
1/200 i-OA	+	<1	148	3.820
1/100 i-Tg + 100 µg/ml Tg	+	25	<1	3.860
1/100 i-Tg + 100 µg/ml OA	+	122	<1	4.180
1/100 i-OA + 100 µg/ml OA	+	<1	15	3.260
1/100 i-OA + 100 µg/ml Tg	+	<1	162	3.880

beads. This response was antigen-specific both in its efferent phase and in its afferent phase (only spots in plates coated with the relevant antigen; strong inhibition by addition of free antigen during the culture). Activation apparently comprises the expression of receptors for growth and differentiation factors, since addition of these factors to the cultures was an absolute requirement for the generation of AFC (not shown in detail), whereas addition of the factors without antigens only resulted in an increase of the IgM "background" AFC.

Apparently, the normal human B cell repertoire contains autoreactive B cells that can be activated by the autoantigen thyroglobulin, if the necessary (soluble) T cell signals are provided. However, at least part of the Tg-reactive B cell repertoire in patients' blood can be driven to (IgG) anti-Tg production by i-Tg only, i.e. without exogenous growth and differentiation factors. We are currently investigating whether this is a unique property of autoreactive B cells.

REFERENCES

1. Defranco, A.L., Raveche, E.S., Asofsky, R. and Paul, W.E. (1982) J.Exp.Med. 155, 1523.
2. Howard, M. and Paul, W.E. (1983) Ann.Rev.Immunol. 1, 307.
3. Katz, D.H., Hamaoka, T. and Benacerraf, B.(1973) J.Exp.Med. 1405.
4. Schimpl, A. and Wecker, W. (1972) Nature New Biol. 237, 15.
5. Nossal, G.J.V. and Pike, B.L. (1980) Proc.Natl.Acad.Sci USA 77, 1602.
6. Logtenberg, T., Jonker, M., Kroon, A., Gmelig-Meyling, F.H.J. and Ballieux, R.E. (1985) Immunol.Lett. 9, 343.
7. Logtenberg, T., Kroon, A., Gmelig-Meyling, F.H.J. and Ballieux, R.E. (1986) J. Immunol., in press.
8. Muraguchi, A. and Fauci, A.S. (1982) J. Immunol., 129, 1104.

ISOTYPE RESTRICTION OF IDIOTOPES FOUND ON HUMAN

ANTI-CARBOHYDRATE ANTIBODIES

F. Emmrich, G. Zenke, and K. Eichmann

Max Planck Institut für Immunbiologie
Stübeweg 51,
D-7800 Freiburg/Br., FRG

SUMMARY

Sharing of idiotopes between IgM and IgG antibodies has been taken as evidence for the integrity of V gene products transferred during isotype switching from μ to γ heavy chains.

Here we describe idiotopes which are exclusively found either on IgM or on IgG antibodies from the same human individual and with the same specificity for the monosaccaride hapten N-acetyl-D-glucosamine (GlcNAc)). Moreover, IgG-restricted idiotopes which occur in high frequency on anti-GlcNAc antibodies of many individuals are also found to be constantly restricted to IgG. Beside the possibility that idiotope composition might be altered more dramatically during isotype switch of anti-GlcNAc B cells compared to other specifities it should be considered that distinct V-genes are used in IgM versus IgG response to GlcNAc.

DISCUSSION

N-acetyl-D-glucosamine (GlcNAc) is the immunodominant haptenic side-chain of the streptococcal group A carbohydrate (A-CHO). Considerable amounts of antibodies against GlcNAc are found in human serum (1), which can be purified by affinity chromatography on GlcNAc-Sepharose 4B (1,2). Serum of a healthy human individual (MSS) was selected, which contained in isoelectric focusing a prominent IgG anti-GlcNAc spectrotype (1A) comprising about 7% of all anti-A-CHO antibodies of this particular donor (3,4). Monoclonal antibodies (mAbs) to purified immunoglobulins were obtained by somatic cell fusion (4,5) and screened for anti-idiotype (anti-Id) antibodies which would bind to anti-GlcNAc but not to anti-GlcNAc depleted antibodies of donor MSS in solid-phase ELISA (5). Both preparations were adjusted to the same IgM/IgG ratio.

Surprisingly, all anti-Id mAbs we have generated until now showed exclusive binding to either IgM anti-GlcNAc or IgG anti-GlcNAc (6). This situation is illustrated in fig.1 for five IgM-restricted versus four IgG-restricted idiotopes. The latter ones (15, 539, 812, 433)[1] are found on the IgG spectrotype 1A of donor MSS (4).

---

[1]The number denotes a distinct idiotope, whereas the full code (for instance 13F15) designates the corresponding mAb.

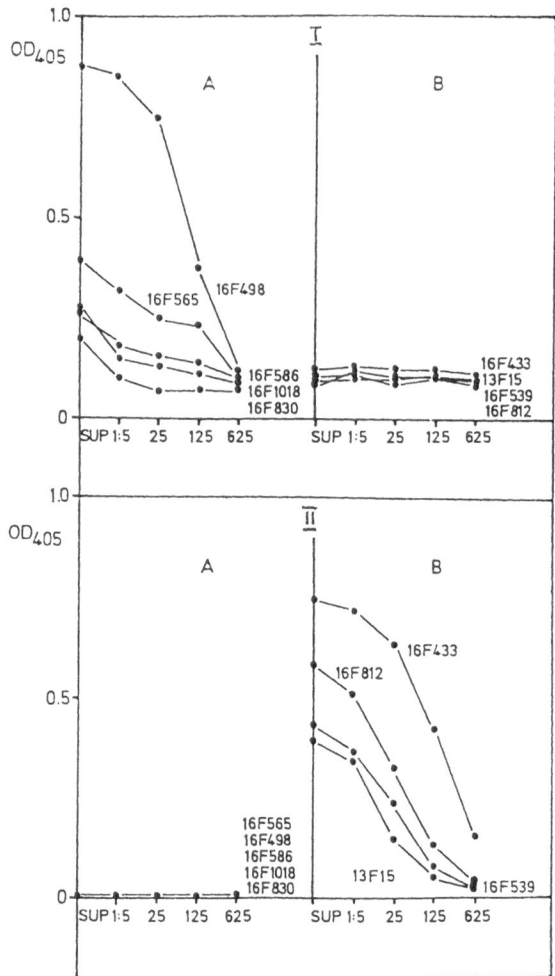

Figure 1

Isotype restriction of idiotopes on anti-GlcNAc antibodies from donor MSS demonstrated in solid-phase ELISA. Equal amounts of affinity-purified anti-GlcNAc antibodies separated by HPLC into the IgM (I) and the IgG (II) fraction were coupled to different ELISA-microtiter plates. After blocking unreacted plastic sites these plates were incubated with culture supernatants (SUP) from anti-Id hybridomas adjusted to 1 µg/ml mouse immunoglobulin (mIg). The anti-Id mAbs (designation is given in the figure) were arranged according to proceding experiments for their isotype-restriction to either IgM (A) or IgG (B). Binding was demonstrated by an enzyme-labelled anti-mIg (for technical details see ref. 4, 6). OD_{405} is the optical density at 405 mm.

Some of the anti-Id mAbs bind different portions of additional IgG anti GlcNAc antibodies of donor MSS with a maximum of 80% of all anti-GlcNAc antibodies for 16F433. Except of 16F539 all anti-Ids were inhibitable by antigen to different degrees, i.e. the corresponding idiotopes are located in or near the antigen combining site. They are distinct from each other and could be mapped into two clusters within the variable region of spectrotype 1A (7). Among IgM-restricted anti-Ids the antibodies 16F565 and 16F489 detect the most frequent idiotopes with precursor B cell frequencies for Id565 of about 10% of IgM anti-A-CHO producing B cells in polyclonally activated MSS PBL cultures.

Even more surprising than the isotpye-restriction in one individual was the observation of exactly the same restriction in unrelated individuals as has been detected with cross-reactive idiotopes 812 and 433 (6).

Possibly, the phenomenon of idiotype/isotype restriction could be interpreted as an initial commitment of human GlcNAc specific B cells for either IgM or IgG production with preferential association of certain V genes with either $C\mu$ or $C\gamma$. In addition, switching from IgM to IgG production should either occur rarely or not at all in these B cells, which may belong to a certain subpopulation. This assumption is difficult to prove experimentally since in general isotype switching is hard to detect in human B cells cultures.

On the other hand, if IgG anti-GlcNAc B cells arose by switching from IgM B cells in vivo, two other alternative interpretations for the isotype restriction might be considered: [1] A large number of different μ-associated anti-GlcNAc variable regions could prevent the detection of a few V regions which might be expanded to large amounts only after switching from μ to γ. However, in a direct binding assay (not shown) a 300-600 fold excess of idiotype-negative IgM specific for GlcNAc would be required to "hide" a clontypic idiotope in specific IgM thus preventing its detection by an IgG-restricted anti-Id (6). This possibility seems unlikely at least for the non-clonotypic idiotopes. [2] The isotype switch could be accompanied by a dramatic change of the idiotype composition. Somatic mutations, accumulated during isotype switching may result in idiotope loss and generation of new idiotopes. Likewise, complex idiotopes which involve parts of the H-chain constant region may disappear or may be newly created. Somatic mutations are more often seen in IgG and IgA than in IgM V_h sequences (8), which might indicate a preferential accumulation of somatic mutations during isotype switching. If so, the IgM anti-GlcNAc repertoire generated by a limited number of germ-line genes might be lost upon switch to IgG, whereas the IgG anti-GlcNAc repertoire is newly generated from mutated V genes not necessarily involved in the primary IgM repertoire.

Regardless of the correct interpretation, it should be noted that idiotype-isotype restriction within IgG subclasses has been described previously also for mouse antibodies (9), and therefore seems to be a more general phenomenon. Studies with other antigenic specificities should provide further information on how often a preferential association of idiotopes and isotype occurs. If this proves to be frequently the case in man it must be given special consideration in future attempts towards the use of anti-Id mAbs for medical purposes such as vaccination or immunosuppression.

REFERENCES

1. Emmrich, F., B. Schilling, and K. Eichmann. J. Exp. Med. 161:547 (1985).
2. Eichmann, K., J. Greenblatt. J. Exp. Med. 133:424 (1971).
3. Emmrich, F., G. Zenke, C. Polke, and K. Eichmann, Ann. Immunol. (Inst. Pasteur) 135C: 95 (1984).
4. Zenke, G., K. Eichmann, and F. Emmrich. Eur. J. Immunol. 14: 164 (1984).
5. Emmrich, F., B. Greger, and K. Eichmann. Eur. J. Immunol. 13: 273 (1983).
6. Emmrich, F., G. Zenke, and K. Eichmann. Submitted.
7. Zenke, G., K. Eichmann, and F. Emmrich. J. Immunol. (December issue) in press (1985).
8. Crews, S., J. Griffin, H. Huang, K. Calame, and L. Hood. Cell 25: 59 (1981).
9. Scott, M.G., and J.B. Fleischmann. J. Immunol. 128: 2622 (1982).

CONTROL OF THE MITOTIC B CELL CYCLE: FACTS AND SPECULATIONS

Fritz Melchers

Basel Institute for Immunology
Grenzacherstrasse 487
CH-4058 Basel, Switzerland

INTRODUCTION

During their differentiation from stem cells and precursors B lympho-cytes reach a stage when they become resting. These resting B lympho-cytes can be activated into cell cycle again when they meet antigen or polyclonal activators. This paper gives a synopsis of the reactions that lead to activation of B cells from the resting state and that control successive B cell cycles.

For the initiation of most B cell responses binding of antigen to immunoglobulin (Ig) on the surface membrane is a necessary but not a sufficient first step. B cell activations that apparently circumvent this binding step to surface Ig are the polyclonal activators lipo-polysaccharide (LPS) (Andersson et al., 1972) and lipoproteins (Melchers et al., 1975). Multiple controls in the process of B cell activation into cell cycle first became apparent when it was found that most antigen-specific B cell responses needed the cooperation of helper T lymphocytes (Claman et al., 1966; Davies et al., 1967; Mitchell and Miller, 1968) and of accessory (A) cells (Mosier, 1967). Many of these cooperating interactions of B cells with T cells and A cells appear to be governed by the membrane proteins encoded in the major histocompatibility complex (MHC), in particular by those of class II molecules (Ia-antigens (Katz et al., 1973; Sprent, 1978).

Not all B cell responses are T cell-dependent. Two types of T cell-independent antigens have been distinguished (Mosier and Subbarao, 1982). Responses to these antigens are not governed by MHC-class II antigens. Finally, Ig-specific antibodies, especially in immobilized forms, will polyclonally activate B cells in T cell-independent, MHC-unrestricted ways (Parker, 1975; Leptin, 1985).

It has recently become evident that activated murine B lymphocytes control activation into the first and into subsequent mitotic cell cycles by interactions with at least three different types of ligands. These ligands are antigen and, what we have called α factors and β factors (Melchers and Lernhardt; Melchers et al., 1985).

BINDING OF ANTIGEN INVOLVES IMMUNOGLOBULIN (Ig) AND, IN SOME CASES, ALSO CLASS II MAJOR HISTOCOMPATIBILITY COMPLEX (MHC) MOLECULES ON THE SURFACE OF B CELLS.

T cell-dependent antigens first bind to surface Ig and, then, may be processed and represented on the surface in the context of class II MHC antigens (Rock et al., 1984; Lanzavecchia, 1985; Tony and Parker, 1985). Helper T cells are thought to then establish contact to antigen and class II MHC molecules on the B cells via their T cell receptor molecules which recognize antigen in the context of class II MHC molecules (Katz et al., 1973; Sprent, 1978). A mosaic of T cell receptors in the T cell membrane may crosslink several surface Ig and class II MHC molecules for efficient triggering.

For helper T cell-independent antigens this binding to the appropriately specific Ig alone is apparently sufficient, although crosslinking of several surface Ig molecules by repetitive determinants of these T-independent antigens appears to be needed for successful signalling to the B cell. In this case possible processing of antigen has not yet been investigated. MHC class II antigens on the surface of B cells appear not to be involved in triggering by T-independent antigens. Ig-specific polyclonal and monoclonal antibodies directed towards the constant region domains of μ-heavy chains (Parker, 1975; Leptin, 1985) and κ-light chains can activate B cells polyclonally, i.e. independent of their specificity for antigen. Crosslinking of surface Ig, achieved by Ig-specific antibodies immobilized on Sepharose beads, is again needed for successful signalling.

Allo-MHC-reactive helper T cells apparently only need to interact with class II MHC molecules on B cells, while polyclonal activators of bacterial origin, such as lipopolysaccharides (LPS) or lipoproteins circumvent both Ig and class II MHC molecules for signalling. However, surface Ig is somehow involved in these stimulatory signals since soluble (Fab)$_2$-fragments of Ig-specific antibodies inhibit stimulation by LPS or lipoprotein (Andersson et al., 1974). It has long been suspected that a functional complex of molecules including Ig, class II MHC molecules, LPS receptors and other receptors are involved in this signalling reaction.

THE PRODUCTION OF α FACTORS

α factors are secreted by normal, activated macrophages and by macrophage lines. There are T-dependent and T-independent ways to activate macrophages (also called accessory (A) cells) to α-factor production. Polyanions such as dextransulfate, LPS and lipoprotein, as well as immune complexes are potent T cell-independent activators of A cells for α factor production (Corbel and Melchers, 1984). Helper T cells recognize processed antigen in the context of class II MHC molecules on A cells and thereby, are expected to activate A cells to α factor production. While a wealth of long-term tissue culture lines of normal and malignantly transformed A cells, T cells and B cells have been found to produce α factors (Corbel and Melchers, 1984) it remains to be clarified whether normal, short-term activated T or B cells can produce and secrete α factors. This appears to be an important issue to settle since B cells have been found to present antigen to helper T cells (Rock et al., 1984; Lanzavecchia, 1985; Tony and Parker, 1985). If neither normal T nor normal B cells could not produce and secrete α factors then their interaction with each other and with antigen would not lead to α-factor-dependent B cell proliferation (see last paragraph).

THE ACTION OF α FACTORS ON B CELL PROLIFERATION CAN BE REPLACED BY CROSS-LINKED C3, C3b AND C3d BUT NOT BY THEIR SOLUBLE COUNTERPARTS

Sepharose-bound or glutaraldehyde-aggregated C3, C3b or C3d replaces α factors in the activation and cell cycle control of B cells (Erdei et al., 1985; Melchers et al., 1985). This indicates that receptors for the complement component C3d (called CR2 in the human system (Fearon, 1984; Ross, 1980)) are involved, and that crosslinking of these receptors is important for stimulation of B cell proliferation. It should be mentioned that CR2 receptors are expressed specifically on B cells, but not on T cells or A cells. Furthermore, monoclonal antibodies specific for the CR2-receptor have been found to stimulate human B cells to proliferation (Frade et al., 1985). The problem arises how soluble α factors from A cells could achieve similar crosslinking of CR2 receptors. The hypothesis may be advanced that microaggregation of components of the early pathway of complement activation could lead to a microaggregation of the CR2 receptors. This early pathway of complement activation may be initiated by the binding of antigen to Ig, which, in turn, activates C1q. This, in turn would first activate C1r and C1s, subsequently C2, C4 and finally C3. CR2, in this view, would be the final receptor site of a successful cascade activation of complement. Macrophages are known to produce most of these complement components, and it remains to be investigated whether B cells (or T cells) also produce some of them.

While insolubilized C3d stimulates B cells along the mitotic cycle, soluble C3d inhibits this as well as the α-factor-mediated stimulation (Melchers et al., 1985). The same CR2 receptor on B cells may, therefore, signal these cells positively or negatively for progression through the cell cycle, depending on whether the receptors are crosslinked in the surface membrane, or whether they interact with single ligands.

THE INTERACTIONS OF β FACTORS WITH MURINE B CELLS INVOLVE SO FAR UNKNOWN RECEPTORS ON B CELLS

β factors appear to be produced by helper T cells when they recognize antigen in the context of class II MHC molecules. A cells as well as B cells may suffice as antigen-presenting cells to induce this antigen-specific, MHC-restricted β factor production of T cells. The molecular nature of β factors remain to be clarified. The highly purified preparations of B151-TRF (Harada et al., 1985; Takatsu et al., 1985) appear to act as β factors (Takatsu, K., Harada, N., Takahashi, T.C. and Melchers, F., manuscript in preparation). BSF-1 (Ohara et al., 1985; Ohara and Paul, 1985) on the other hand, appears to be an early acting factor with no detectable β factor activity (Ohara, Paul and Melchers, to be published). It is remarkable that even the most purified B151-TRF preparations have been reported to stimulate activated, A-cell containing B cells populations both to proliferation and to maturation into Ig-secreting cells, while they stimulate resting B cells to maturation only (Harada et al., 1985; Takatsu et al., 1985; Melchers et al., 1980; Oliver et al., 1985; Leclercq et al., 1985). The experiments with activated B cells have now been repeated with A cell-depleted B cell populations. We have found that in this absence of any source of α factors the B cell blasts did mature but did not proliferate any further. Therefore, it appears that β factors induce maturation and proliferation in synergy with α factors, but maturation only in the absence of α factors.

There are antigen-unspecific, polyclonal ways in which β factors can either be produced or be replaced. Plant agglutinins such as Concanavalin A can stimulate helper T cells in A cell-independent ways to β-factor production. However, no physiologically relevant way is known for such A- (or B-) cell-independent T cell stimulation. Polyclonal B cell activators such as LPS or lipoprotein replace β factors in their action on B cell proliferation. When they stimulate B cells to proliferation they, therefore, replace the action via Ig as well as via β-factor-receptors, but need the costimulatory action of α factors (Melchers, F., to be published).

THE THREE TYPES OF LIGANDS, I.E. ANTIGEN, α FACTORS AND β FACTORS, CONTROL THE MITOTIC CYCLE OF ACTIVATED B CELLS AT THREE DIFFERENT RESTRICTION POINTS

Studies on the control of the cell cycle of activated B cells have been facilitated by our ability to synchronize the activated cells by size selection using velocity sedimentation (Miller and Phillips, 1969; Melchers and Lernhardt, 1985). The first restriction point occurs directly after mitosis and is controlled by the occupancy of surface Ig. Experiments using Sepharose-bound Ig-specific monoclonal antibodies have shown that surface Ig may be occupied for as short as 15 minutes (and as long as 36 hours) to prepare the cells for the next interaction.

The second restriction point occurs some 4 hours into the 20 hour long cell cycle, within the G1 phase of the cell cycle. It is controlled by α factors. B cell will not enter S phase, i.e. DNA replication, unless α factors are present. This makes α-factors, and consequently C3d, similar in action to insulin-like growth factors. It is, therefore, interesting to note that a very weak, but significant homology in nucleotide sequence has been found in the C3d part of the α-chain of C3 when compared to murine epidermal or nerve growth factors (Melchers et al., 1985).

The third restriction point occurs 2 to 4 hours before mitosis and is controlled by β factors. It is worth remembering that it is at this late point into the G2 phase of the cell cycle where sister chromatids of recently replicated chromosomal DNA exist which could undergo sister chromatid exchange. Sister chromatid exchange has been implied as a mechanism for Ig class switching (Honjo et al. 1981). This, in turn, has been suggested to be controlled by T cells. A prolonged stay at this point in G2 effected by temporary lack of T cell help, followed by rescue with renewed help, would increase the probability for sister chromatid exchanges and thereby, maybe, for class switching.

It remains to be emphasized how remarkable the synchrony of dividing B cells remains for more than five consecutive cell cycles. B cell lymphomas and myelomas never display such synchrony. The hypothesis has been advanced that synchrony in all activated B cells of a population of a given size is maintained by the secretion of a division-controlling mediator which is made, and also sensed, by all B cells (Melchers and Lernhardt, 1985). This may occur at a late stage in the cell cycle, shortly before mitosis. It could be analogous to the control of synchronized aggregation competence of Dictyostelium by the secretion and reception of cyclic AMP (Melchers and Lernhardt, 1985). If such a mechanism for the control of mitosis of B cells exists it may explain the growth of foci of B cells in vivo. B cells would preferentially divide in areas which are high in concentrations for such mitosis-mediating substances, i.e. in areas where other activated B cells divide. It may also explain why normal B cells cease to divide in vitro below a critical cell concentration, i.e. 10^4 cells/ml (Andersson et al., 1977). Less than 10^4 B cells may not release sufficient amounts of these mitosis-mediating substances into one milliliter of tissue culture medium.

In summary, activated B cells control their mitotic cell cycle by three interactions with their environment at three points during the cell cycle. Antigen acts immediately after mitosis, excites the cells to enter the G1 phase up to the point, where macrophages (A cells) have to contribute α factors so that B cells can enter S phase. In the prolonged absence of α factors B cells appear to mature to Ig secreting cells and, thereby, may loose the capacity to enter S phase. When B cells have traversed S phase in the presence of α factors, they then need β factors produced by helper T cells to enter mitosis.

DEREGULATIONS OF THE B CELL CYCLE CONTROL ARE EXPECTED TO LEAD TO IMMUNODEFICIENCES, TO AUTOIMMUNITIES, TRANSFORMED GROWTH AND FINALLY, TO MALIGNANCY.

For progression through the mitotic cycle three types of ligands are required (r) at three restriction points in the cell cycle of normal B cells. These functions of a B cell may change so that signalling at a given point either becomes defective (d), or that signalling will become independent (i) of the ligand. In the former change we may expect that ligand binding either no longer occurs or does not lead to stimulation. A change to deficient signalling at any of the three restriction points (anti-Ig^r → anti-Ig^d, or α^r → α^d, or β^r → β^d) should lead to B cell immunodeficiencies and, thus, to a defect in humoral responses. In the latter change B cells will proceed beyond a given restriction point despite the absence of an interaction with the appropriate ligand, or because of a deregulated, excessive presence of the ligand. Changes from r to i may be considered oncogenic events leading to partial transformation of B cells. Full transformation would be expected whenever a B cell has become independent for stimulation at all three restriction points (anti-Ig^i, α^i, β^i).

Changes to ligand-independent signalling at the first restriction point only (anti-Ig^i) are expected to result in polyclonal activation of B cells to proliferation and Ig secretion, whenever α- and β-factors are provided by the environment. The potent recognition of any foreign antigen presented by A cells to antigen-specific helper T cells that leads to such α- and β-factor production. should suffice. If the repertoire of B cells also contains autoreactive B cells antibody production and systemic autoimmunity should be induced.

Changes to ligand-independent signalling at the second (α^i) or third (β^i) restriction point should lead to an increased susceptibility of B cells to either A cell-<u>independent</u>, T cell-<u>dependent</u>, or to T cell-<u>independent</u>, A cell-<u>dependent</u>, antigen-specific, i.e. to oligoclonal B cell stimulation. In the A cell-independent, T cell-dependent situation autoimmune B cells may be stimulated whenever autoimmune T cells exist in the system, while in the T cell-independently A cell-dependent situation self antigens must stimulate A cells to α-factor production for a successful autoantibody production.

Whenever changes to ligand-independent signalling occur at two restriction points in one B cell, autoimmune B cells should be stimulated as easily as those which react to foreign antigens. Ig^i, α^i-B cells only should need helper T cells to recognize any antigen in the context of MHC class II antigens on any cells (regardless of whether this cell can produce α factors or not), and this, of course, includes the antigen that such B cells bind themselves. The resulting β-factor production by the T cells would suffice for polyclonal B cell activation. Ig^i, β^i-B cells should be induced to polyclonal stimulation wherever they meet an A cell that has been stimulated to α-factor production. Viral and, particularly for the latter case, bacterial infections appear to be potent activators in such situations.

Finally, α^i, β^i-B cells will be stimulated by any antigen alone, and even in the absence of A or T cell-derived α- and/or β-factor production, as long as the requirements for stimulation via surface Ig, i.e. possibly crosslinking of Ig receptors by repetitive determinants of the antigen, are met.

The changes in the requirements for cell cycle progression can occur during the somatic generation of B cells and their stimulation to Ig-secreting plasma cells. That may lead to systemic autoimmune disease whenever signalling via Ig becomes ligand independent, and whenever more than one restriction point is changed to ligand-independent signalling. On the other hand, whenever single signalling via α or via β factors becomes ligand-independent local autoimmunity and autoimmune disease at specific sites in the body may develop from a change in an autoimmune B cell with specificity for a given local autoantigen. These auto-immunities should be either T cell dependent, A cell independent or A cell-dependent, T cell independent. Switches in the classes of Ig and somatic mutations in Ig genes leading to better fitting antibodies are expected to increase the severity of the autoimmune diseases.

Changes in signalling could, on the other hand, also occur in the germline, i.e. become heritable. Several strains of mice exist which display hyperreactivity and hyperplasia of B cells with the development of increased numbers of Ig-secreting cells and autoimmunity, and which, therefore, appear to be good candidates for carrying such changes. We have recently begun to analyze the growth requirements of the LPS-activable B cell populations of several of these mouse strains. First results indicate that activated B cells from NZW mice are Ig^i, α^r, β^r, from NZB mice Ig^r, α^r, β^i and from BxSB mice Ig^r, α^i, β^i while those from control C57BL/6J nu/nu mice are, as expected, Ig^r, α^r, β^r (Melchers, 1985). These results indicate that it should be possible to breed mouse strains with different combinations of r and i phenotypes for signalling with anti-Ig, α and β factors, if the three phenotypes are encoded on three different chromosomes. It should then be possible to test the influences of these changes on the onset and development of B cell hyperplasia, autoimmunity, transformation and malignancy.

B CELLS BALANCE BETWEEN PROLIFERATION AND MATURATION

It has long been known that the earliest reactions of B cells after activation lead to maturation to Ig secretion, in time well before these B cells enter S phase (Melchers and Andersson, 1979). With increasing numbers of division B cells increase the rate of Ig secretion when compared to the rate of total protein synthesis in these cells (Melchers and Andersson, 1974). Maturation to Ig secretion in subsequent cell cycles occurs in the G1 phase of the cell cycle. This is evident from measurements of the rate of synthesis of secretory-type IgM, from the expression of secretory-type μ heavy chain mRNA and from the occurence of IgM-secreting, plaque-forming cells during the cell cycle of synchro-nized, LPS-stimulated, activated B cells. β factors appear to be needed to stimulate Ig secretion in G1, as it has been shown with resting B cells (Melchers et al., 1980), and as it is evident from the cell cycle control experiments, which show β factors to control mitosis of activated B cells and, therefore, their entry into the G1 phase (Melchers and Lernhardt, 1985). Excitation of resting B cells by Ig-specific anti-bodies alone, on the other hand, appears to be insufficient to induce maturation to Ig secretion, but rather prepares the cells for subsequent susceptibility to α-, and then to β-factor action for the next cell cycle.

When activated B cells have been exposed to β-factor action and have divided, and when they have then even been excited by Ig-specific anti-bodies, they will remain in G1 and <u>not</u> enter S phase when α-factors are <u>not</u> provided. The balance of an activated B cell between proliferation and maturation, therefore, appears to be controlled by the presence or absence of α factors which control the entry into S phase.

Maturation to Ig secretion in the absence of proliferation leads to an abrogation of the capacity of such B cells to be induced to pro-liferation in a later stimulation. Thus, in 48 hours 99% of the maturing B cells have lost the capacity to be induced for proliferation (Melchers et al., 1980). In the absence of α factors antigen-specific B cell responses are, therefore, expected to be eliminated rather than stimulated.

ELIMINATION OF SELF REACTIVE B CELLS

If normal helper T cells and B cells, maybe at a given stage of development from stem cells, cannot produce α factors, and if they interact in an A cell-deficient environment, their MHC-restricted, antigen-specific interaction will lead to excitation of the B cells and to β-factor production by T cells. In the absence of α-factors this will lead to maturation without proliferation of the B cells and effectively, to an elimination of these antigen-specific B cells. This abortive interaction could be a mechanism by which the B cell compartment of the immune system filters its repertoire for self-antigen-reactive B cells. If the corresponding self-antigen-reactive (MHC-restricted?) T cells exist in the system, this could lead to B cell tolerance to self.

ACKNOWLEDGEMENTS

The Basel Institute for Immunology was founded and is supported by F. Hoffmann-La Roche, Limited Company, Basel, Switzerland.

REFERENCES

Andersson, J., Sjöberg, O. and Möller, G. 1972, Induction of immuno-globulin and antibody synthesis in vitro by lipopolysacchardies, Eur. J. Immunol. 2:349-353.
Andersson, J., Bullock, W.W. and Melchers, F. 1974, Inhibition of mito-genic stimulation of mouse lymphocytes by anti-mouse immunoglobulin antibodies. I. Mode of action, Eur. J. Immunol. 4:715-718.
Andersson, J., Coutinho, A., Lernhardt, W. and Melchers, F. 1977, Clonal growth and maturation to immunoglobulin secretion in vitro of every growth-inducible B-ylmphocyte. Cell 10:27-34.
Claman, H.N., Chaperon, E.A. and Triplett, R.F. 1966, Thymus-marrow cell combinations. Synergism in antibody production, Soc. Exp. Biol. & Med. 122:1167-1171.
Corbel, C. and Melchers, F. 1984, The synergism of accessory cells and of soluble α-factors derived from them in the activation of B cells to proliferation. Immunol. Rev. 78:51-74.
Davies, A.J.S., Leuchars, E., Wallis, V., Marchant, R. and Elliott, E.V. 1967, The failure of thymus-derived cells to produce antibody. Transplantation 5:222-230.
Erdei, A., Melchers, F., Schulz, T. and Dierich, M. 1985, The action of human C3 in soluble or cross-linked form with resting and activated murine B lymphocytes. Eur. J. Immunol. 15:184-188.
Frade, R., Crevon, M.C., Barel, M., Vazquez, A., Krikorian, L., Charriaut, C. and Galanaud, P. 1985, Enhancement of human B cell proliferation by an antibody to the C3d receptor, the gp 140 molecule. Eur. J. Immunol. 15:73-76.

Harada, N., Kikuchi, Y., Tominaga, A., Takaki, S. and Takatsu, K. 1985, BCGFII activity on activated B cells of a purified murine T cell-replacing factors (TRF). J. Immunol. 130: No. 5 2219-2224.

Honjo, T., Nakai, S., Nishida, Y., Kataoka, T., Yamawaki-Kataoka, Y., Takhashi, N., obata, M., Shimiuzu, A., Yaoita, Y., Nikaido, T. and Ishida, N. 1981. Immunol. Rev. 59:5-32.

Iscove, N.N. and Melchers, F. 1978, Complete replacement of serum by albumin, transferrin, iron and soybean lipid in cultures of lipopoly-saccharide-activated B lymphocytes. J. Exp. Med. 147:923-933.

Katz, D.H., Hamaoka, T., Dorf, M.E. and Benacerraf, B. 1973, Cell interactions between histocompatible T and B lymphocytes. III. Demonstration that the H2-gene complex determines successful physiologic lymphocyte interactions. Proc. Natl. Sci. USA 70:2624-2628.

Lanzavecchia, A. 1985; Antigen-specific interaction between T and B cells. Nature 314:537-539.

Leclercq, L., Cambier, J.C., Mishal, Z., Julius, M.H. and Theze, J. 1985, Supernatant from a cloned helper T cell stimulates most small resting B cells to undergo increased I-A expression, blastogenesis and progression through cell cycle. Submitted.

Leptin, M., Potash, M.H., Grützmann, R., Heusser, C., Schulman, M., Köhler, G. and Melchers, F. 1984, Monoclonal antibodies specific for murine IgM I. Characterization of antigenic determinants on the four constant domains of the μ-heavy chain. Eur. J. Immunol. 14,534-542.

Leptin, M. 1985, Monoclonal antibodies specific for murine IgM. II. Activation of B lymphocytes by monoclonal antibodies specific for the four constant domains of IgM. Eur. J. Immunol. 15:131-137.

Melchers, F. 1985, Murine B cell cycle control and its deregulations in NZW. NZB and BXSB mice. In: Proceedings of a Workshop on Primary Immunodeficiency Diseases, Gmunden Austria, Elsevier Amsterdam (in press).

Melchers and Andersson. 1974a, Early changes in immunoglobulin M synthesis after mitogenic stimulation of bone marrow-derived lymphocytes. Biochemistry 13:4645-4653.

Melchers, F., Andersson, J. 1974b, Changes in synthesis, turnover and secretion, and in numbers of molecules on the surface of B cells after mitogenic stimulation. Eur. J. Immunol. 4:181-188.

Melchers, F., Andersson, J., Lernhardt, W. and Schreier, M.H. 1980, B cells induced by antigen-activated T cell help factors, Eur. J. Immunol. 10:669-685.

Melchers, F., Braun, D. and Galanos, C. 1975, The lipoprotein of the outer membrane of Escherichia coli: A B-lymphocyte mitogen, J. Exp. Med. 142:473-482.

Melchers, F., Corbel, C., Leptin, M. and Lernhardt, W. 1985, Activation and cell cycle control of murine B lymphocytes. J. Cell. Sci. Suppl. 3 (in press).

Melchers, F., Erdei, A., Schulz, T and Dierich, M. 1985, Growth control of activated, synchronized murine B cells by the C3d fragment of human complement. Nature 317:264-267.

Melchers, F., Lernhardt, W. 1985, Three restriction points in the cell cycle of activated murine B lymphocytes, Proc. Natl. Acad. Sci. USA (in press).

Miller, R.G. and Phillips, R.A. 1969, Separation of cells by velocity sedimentation. J. Cell. Physiol. 73:191-202.

Mitchell, G.F. and Miller, J.F.A.P. 1968, Cell to cell interaction in the immune response. II. The source of hemolysin-forming cells in irradiated mice given bone marrow and thymus or thoracic duct lymphocytes. J. Exp. Med. 128:821-837.

Mosier, D.E. 1967, A requirement for two cell types for antibody formation in vitro. Science 158:1573-1574.

Mosier, D.H. and Subbarao, B. 1982, Thymus-independent antigens:

complexity of B lymphocyte activation revealed. Immunol. Today 3:217-222.

Ohara, J. and Paul, W.E. 1985, Production of a monoclonal antibody to and molecular characterization of B-cell stimulatory factor-1, Nature 315:333-335.

Ohara, J., Lahet, S., Inmann, J. and Paul, W.E. 1985, Partial purification of murine B-cell stimulatory factor (BSF)-1. J. Immunol. (in press).

Oliver, K., Noelle, R.J., Uhr, J.W., Krammer, P.H. and Vitetta, E.S. 1985, B cell growth factor (BCGF I or BSPp1) is a differentiation factor for resting B cells and may not induce cell growth. Proc. Natl. Acad. Sci. USA. (in press).

Parker, D.C. 1975, Stimulation of mouse lymphocytes by insoluble anti-mouse immunoglobulins. Nature 258:365.

Rock, K.L., Benacerraf, B. and Abbas, A.K. 1984, Antigen presentation by hapten-specific B lymphocytes. I. Role of surface immunoglobulin receptors. J. Exp. Med. 160:1102-1113.

Ross, G.D. 1980, Analysis of the different types of leukocyte membrane complement receptors and their interaction with the complement system. J. Immunol. Meth. 37:197-211.

Sprent, J. 1978, Role of H-2 gene products in the function of T helper cells from normal and chimeric mice measured in vivo. Immunol. Rev. 42:108-137.

Takatsu, K., Harada, N., Hara, Y., Takahama, Y., Yamada, G., Dobashi, K. and Hamaoka, T. 1985, Purification and physicochemical characterization of murine T cell replacing factor (TRF), J. Immunol. 134:382-388.

Tony, H.-P. and Parker, D.C. 1985, Major histocompatibility complex-restricted, polyclonal B cell responses resulting from helper T cell recognition of anti immunoglobulin presented by small B lymphocytes. J. Exp. Med. 161:223-241.

THE EFFECTS OF INTERLEUKIN 2, GAMMA INTERFERON, AND B CELL DIFFERENTIATION FACTOR ON THE DIFFERENTIATION OF HUMAN B CELLS

N. Nakagawa, T. Nakagawa, D. J. Volkman, H. Goldstein,
J. L. Ambrus Jr., and A. S. Fauci

LIR, NIAID, and NIH
Bethesda, MD, USA

It has been well known that in the human system, SAC-activated B cells proliferate and differentiate in response to helper T cell factors. Among many factors included within the broad category of helper T cell factors, three distinct factors (interleukin 2 {IL-2}, gamma interferon {γ-IFN}, and B cell differentiation factor {BCDF}) were reported to be involved in B cell differentiation. Recent progress in recombinant DNA technology has made highly purified IL-2 and γ-IFN available, and even though BCDF is not yet cloned by molecular biological techniques, two kinds of BCDFs derived from the supernatants of newly established cell lines have been recently reported. In the present study, employing normal B cells or EBV-transformed B cell lines, we have investigated the effects of these factors on human B cells.

1. STAPHYLOCOCCUS AUREUS COWAN I (SAC) ACTIVATED NORMAL HUMAN B CELLS

Purified B cells were obtained from human peripheral blood by depleting AET-rosette positive cells and plastic adherent cells; in addition leu-1 positive cells were killed by treatment with the respective antibody and complement. These B cells were activated with SAC for 3 days and fractionated into Tac-Ag(+) and Tac-Ag(-) populations by cell sorter analysis (EPICS V). Both unfractionated and fractionated cells were further cultured with or without factors for 3 days and immuno globulin producing cells were detected by a reverse hemolytic plaque forming cell (PFC) assay. Mixed lymphocyte reaction culture supernatant (MLR-CS) was employed as a source of conventional T cell factor. It was concentrated 10 times and added at 10% v/v. IL-2 was used at 50 u/ml and γ-IFN was used at 7 u/ml. These concentrations were used since they were the same as those contained in original MLR-CS. It was observed that IL-2 stimulated only Tac-Ag(+) cells to differentiate, and even though γ-IFN itself did not induce differentiation, it enhanced the effect of IL-2. These effects were completely blocked by anti-Tac antibody. MLR-CS also stimulated only Tac-Ag(+) fractions to differentiate and its effect was almost completely inhibited by anti-Tac antibody. Furthermore, the number of PFCs which were induced by the combination of IL-2 and γ-IFN was approximately equal to that induced by MLR-CS. These data suggested that IL-2 and γ-IFN are the main factors which induce B cell differentiation soon after SAC-activation.

Next, we investigated the kinetics of Tac-Ag expression on the B cell surface following SAC-activation. B blast cells obtained from unfractionated SAC-stimulated (3 days) B cells by Percoll gradient centrifugation, were more than 80% Tac-Ag positive and they behaved similarly to purified Tac-Ag(+) cell fractions. However, when B blast cells were cultured for an additional 4 days without any stimuli, they gradually lost Tac-Ag, and on the 4th day their Tac-Ag positivity was only 30-40%. Therefore, we obtained viable B cells at this point by Ficoll-Paque centrifugation and fractionated these B cells into Tac-Ag(+) and Tac-Ag(-) fractions. Both unfractionated and fractionated cells were cultured with or without factors (IL-2, γ-IFN and MLR-CS) for an additional 3 days and immunoglobulin producing cells were enumerated by the PFC assay. The results were completely different from those of B cells obtained soon after SAC-activation. At this stage, MLR-CS affected unfractionated B cells and both Tac-Ag(+) and Tac-Ag(-) fractions to the same degree, and anti-Tac antibody showed no inhibitory effect. Moreover, no stimulatory effect was observed by IL-2 or the combination of IL-2 and γ-IFN. Therefore, it was more likely that another factor included in MLR-CS (probably BCDF) affected the differentiation of the cells at this later stage of activation. In order ot investigate this possibility, we studied separately 2 populations of activated B cells. One population was obtained after 3 days of SAC stimulation and were referred to as "earlier stage" B cells. Another population was stimulated for 3 days with SAC and then was cultured without factors for an additional 4 days and were referred to as "later stage" B cells. Tac-Ag(-) cells in the earlier stage did not differentiate at all even when factors were added, while Tac-Ag(-) cells in the later stage differentiated in response to MLR-CS.

We next determined the effects of 2 separate BCDFs on the differentiation of B cells at these stages. One BCDF was BCDF-Nal which was recently reported to be purified by Hirano et al. and the other was BCDF-YA2 which was also recently reported by Goldstein et al. in our laboratory. It was clear that IL-2 and γ-IFN manifested effects on B cells only in the earlier stage, while BCDFs affected the cells in the later stage. However, only IL-2 was observed to induce proliferation of cells in both stages, while BCDF had no effect on proliferation at either stage. Therefore, in the earlier stage after SAC activation, IL-2 and γ-IFN serve as the main factors for B cell differentiation, and in the later stage, BCDF is the major factor for differentiation. However, IL-2 and not BCDF affects B cell proliferation in both stages.

2. POKEWEED MITOGEN (PWM) ACTIVATED HUMAN MONONUCLEAR CELLS (MNC)

As has been reported by many authors, when human MNC are stimulated with PWM, B cell differentiation is observed. Since unfractionated MNC were stimulated with PWM in those systems, various factors were secreted from T cells as well as macrophages. Thus, we used cyclosporin A (CsA) to inhibit secretion of several factors including IL-2 from T cells. CsA has been successfully used as an immunosuppressive agent particularly for patients who have undergone organ transplantation. CsA inhibits IL-2 secretion from T cells and thus impaires T cell function. γ-IFN secretion was also reported to be inhibited by CsA. We stimulated purified T cells with phytohemaggulutinin (PHA) (1 µg/ml) for 2 days, and harvested the supernatant. Factor activities in the supernatant were examined. IL-2 activity was measured by proliferation of an IL-2 dependent murine cell line, HT-2. BCDF activity was measured in two separate assays. One was the development of IgG-PFC by CESS and the other was IgM secretion by SKw6.4. Both CESS and Skw6.4 are EBV transformed cell lines and differentiate into immunoglobulin secreting cells in response to BCDF. The result was that IL-2 activity was completely blocked by 1 µg/ml of CsA, but no influence was observed on BCDF activity detected by two separate assays. Moreover, the fractionation of PHA-supernatant demonstrated that BCDF secreted from CsA-treated T cells had the same molecular weight as that secreted from nontreated T cells. Thus despite inhibiting IL-2 production, CsA treatment did not appear to affect either functional or molecular weight characteristics of secreted BCDF.

The same result, that CsA inhibited only IL-2 secretion but not BCDF secretion, was also observed in PWM-stimulated MNC supernatant. However, even though BCDF was demonstrated to be present in cultures containing CsA, it has been reported that CsA completely inhibited immunoglobulin production in the PWM-stimulated MNC system. In this regard, we added back exogenous IL-2 to PWM-stimulated MNC cultures and observed that it substantially reconstituted immunoglobulin secretion which had been blocked by CsA. This result suggested that IL-2 is the main factor which is responsible for immunoglobulin secretion in PWM-stimulated cultures. Next, we stimulated MNC with PWM for 3 days, and separated surface immunoglobulin positive cells by cell sorting (EPICS V). We further cultured them with or without factors (IL-2, BCDF and MLR-CS) for 3 days and enumerated PFC. The results were clear, in that IL-2 had almost the same effect as that of MLR-CS, while BCDF had little effect on PFC responses. Therefore, we concluded that PWM stimulates B cells up to the stage that they can respond to IL-2 by differentiation, but not up to the stage where they respond to BCDF.

3. CB CELLS

We next determined the effects of these factors on the EBV-transformed B cell lines, CB-1 and CB-2. CB-1 was established from a normal individual in our laboratory , and is comprised of 30-40% Tac-Ag(+) cells. CB-1 secretes IgG and when cultured with MLR-CS for 3 days the number of PFC was 5-6 times as many as that of background. Although IL-2 alone induced far fewer PFC in CB-1 than did MLR-CS, and although γ-IFN alone did not enhance background PFC, a synergistic effect was observed when the combination of IL-2 (100 u/ml) and γ-IFN (10 u/ml) was added to the culture; the number of PFC was increased to the same level as that observed with MLR-CS. Next, the kinetics of the synergistic effect of IL-2 and γ-IFN was analyzed. As mentioned above, when IL-2 and γ-IFN were added at the beginning of the culture, they exerted a synergistic effect on differentiation. When γ-IFN was added on the second day, synergy was still observed, but when γ-IFN was added on the third day, it had no effect on differentiation. In contrast, if IL-2 was added either on the second day or on the third day of culture, no synergy was observed. Therefore, it was concluded that in the differentiation of CB, IL-2 and γ-IFN work sequentially: IL-2 delivers the first signal followed by the γ-IFN signal as late as 24 hours after. To our knowledge, CB-1 is the first cell line that has been reported to show this phenomenon. We cultured CB-1 cells for 10 months and obtained CB-2 which had less than 5% Tac-Ag(+) cells. The observation of marker reduction in IL-2 receptor expression after we cultured CB-1 for 10 months is compatible with the reported changes of other B cell surface markers in EBV-transformed normal B cell lines during long term culture (Ref. Nisson and Klein). We feel that CB-1 gradually proceeded along its maturation process, and changed its characteristics with regard to Tac-Ag positivity giving rise to CB-2 with a much lower proportion of Tac-Ag positive cells. We cultured both CB-1 and CB-2 with or without factors (IL-2, γ-IFN, BCDFs and MLR-CS) for 3 days and enumerated IgG producing cells by the PFC assay. For the differentiation of CB-1 cells, the combination of IL-2 and γ-IFN had the same effect as that of MLR-CS, while BCDFs had little effect. However, for differentiation of

CB-2, IL-2 and γ-IFN had no effect, while BCDFs had almost the same effect as that of MLR-CS. Therefore, we conclude that CB-1 represents B cells in the ealier stage after activation, and CB-2 represent B cells in the later stage. Thus, with these B cell lines, we have added weight to our hypothesis that B cells, when activated, gain Tac-Ag, respond to IL-2 and γ-IFN by differentiating, then lose Tac-Ag in the more mature stage and become responsive to BCDF.

CONCLUSIONS

In B cell differentiation, responsiveness to factor changes as B cells proceed along the maturation process. In the earlier stage of activation, they respond to IL-2 and γ-IFN; in contrast, in the later stage of activation, they respond to BCDF.

ACKNOWLEDGEMENTS

We thank Dr. T. A. Waldmann, NIH, USA, for providing us with anti-Tac antibody; Dr. T. Kishimoto and Dr. T. Hirano, Osaka University, Japan, for providing us with purified BCDF; Mr. A. Palini and Dr. M. J. Waxdal for technical assistance in flow cytofluorometry work and Ms. Ann C. London for excellent editorial assistance.

REFERENCES

1. Nakagawa, T., T. Hirano, N. Nakagawa, K. Yoshizaki, and T. Kishimoto. 1985. Effect of recombinant IL-2 and γ-IFN on proliferation and differentiation of human B cells. J. Immunol. 134:959.
2. Muraguchi, A., J. H. Kehrl, D. L. Longo, D. J. Volkman, and A. S. Fauci. 1985. Interleukin 2 receptors on human B cells. Implications for the role of interleukin 2 in human B cell function. J. Exp. Med. 161:181.
3. Hirano, T., T. Taga, N. Nakano, K. Yasukawa, S. Kashiwamura, K. Shimizu, K. Nakajima. K. H. Pyun, and T. Kishimoto. 1985. Purification to homogeneity and characterization of human B-cell differentiation factor (BCDF or BSFp-2). Proc. Natl. Acad. Sci. U.S.A. 82:5490.
4. Goldstein, H., D. J. Volkman, J. L. Ambrus Jr., and A. S. Fauci. 1985. Chracterization of a T4$^+$/Leu8$^+$ T cell clone that directly helps B cell Ig production by secreting B cell differentiation factor. J. Immunol. 135:339.
5 . Nakagawa, T., N. Nakagawa, D. J. Volkman, and A. S. Fauci. 1986. Sequential synergistic effect of interleukin 2 and interferon-γ on the differentiation of a Tac-antigen$^+$ B cell line. J. Immunol. in press.
6. Nilsson, K., and G. Klein. 1982. Phenotypic and cytogenic characteristics of human B lymphoid cell lines and their relevance for the etiology of Burkitt's lymphoma. Adv. Cancer Res. 37:310.

REGULATION OF HUMAN B CELL PROLIFERATION AND DIFFERENTIATION

BY INTERLEUKIN 2

Diane F. Jelinek and Peter E. Lipsky

University of Texas Health Science Center
Southwestern Medical School
5323 Harry Hines Boulevard, Dallas, TX 75235, USA

INTRODUCTION

Interleukin 2 (IL-2) was initially identified as a T cell-derived lymphokine that specifically maintains the growth of activated T cells. Recently, however, it has been demonstrated that B cells can also be induced to express receptors for IL-2 (1). Moreover, a number of functional studies have suggested that IL-2 might play a direct role in B cell activation (2,3). This contention remains controversial however since other reports have suggested that the effect of IL-2 on B cells is an indirect one resulting from an action on contaminating T cells (4).

The current studies were undertaken to investigate in greater detail the role of IL-2 in the activation, proliferation and differentiation of human peripheral blood B cells. In addition to assessing the ability of IL-2 to support B cell responses when present continuously throughout culture, its ability to transmit signals during initial activation and subsequent growth and differentiation was also examined. The results show that IL-2 has direct actions on the B cell, conveying both an early signal that promotes subsequent growth and differentiation and later effects sustaining proliferation and triggering the generation of immunoglobulin-secreting cells (ISC). IL-2 therefore serves as a pleiotropic B cell growth and differentiation factor that alone can support maximal responses of human B cells.

MATERIALS AND METHODS

Cell Separation. Peripheral blood B cells were prepared from peripheral blood mononuclear cells treated with 5mM L-leucine methyl ester as previously described (5). The resultant population of B cells contained approximately 2-3% esterase positive Mϕ, <1% T cells as determined by staining with OKT3 and OKT11 pan T cell monoclonal antibodies, <1% Leu 11b positive NK cells, and >90% B1 positive B cells. In some experiments the fluorescence activated cell sorter (FACS) was used to deplete the B cells further of any contaminating T cells, Mϕ, or NK cells. To accomplish this, B cells were reacted with saturating concentrations of OKT3, OKT11, 63D3 (Mϕ specific) and HNK-1 (NK cell specific), counterstained with fluorescein isothiocyanate-conjugated goat anti-mouse Ig, and the fluorescence negative cells collected using the FACS.

Lymphokine Preparations. For the generation of T cell supernatants (T supt), T cells were stimulated with 1 µg/ml PHA and 1 ng/ml PMA for 2 hr at 37°C. Afterward, the cells were washed 3 times with medium and recultured for an additional 48 hr at 37° C at which time the cell-free culture supernatants were collected. Recombinant purified IL-2 (r-IL-2) and r-IFN-γ were obtained from Amgen Biologicals, Thousand Oaks, CA.

Cell Cultures and Assays of B Cell Responses. For two-stage culture experiments, B cells were stimulated with formalinized Cowan I strain Staphylococcus aureus (SA) organisms in the presence or absence of various T cell lymphokines for 48 hr, washed extensively and recultured in U bottom microtiter wells at 2.5×10^4 cells in 0.2 ml of medium. In other experiments, fresh B cells were cultured in a similar manner without the preincubation. ISC generation detected by using a reverse hemolytic plaque assay and DNA synthesis assayed by the incorporation of [^{3}H]thymidine was determined after a total incubation of 5 days.

RESULTS

IL-2 and T supt support B cell responses. T supt or 100 U/ml r-IL-2 was found to augment DNA synthesis of SA-stimulated B cells (Table I). In similar experiments, either T supt or r-IL-2 also supported the generation of ISC. Similar results were obtained with highly purified B cells obtained by negative selection with the FACS (data not shown).

Table I.   r-IL-2 promotes DNA synthesis and differentiation into ISC of SA-stimulated B cells.

Assay	Lymphokine	Mitogen	
		None	S. aureus
		(cpm × 10^{-3})	
B cell DNA	0	0.7 ± 0.0	0.7 ± 0.0
Synthesis	T Supt	0.5 ± 0.1	17.6 ± 1.8
	r-IL-2	0.5 ± 0.0	12.0 ± 0.6
		(ISC per 10^3 B cells)	
Generation	0	0	0
of ISC	T Supt	1.0	75.6
	r-IL-2	0	78.2

B cells activated for 48 hrs with SA generated large numbers of ISC upon subsequent incubation with T supt (Table II). Without T supt, few ISC differentiated. Experiments were carried out to determine whether IL-2 could also promote responses of SA-activated B cells. As can be seen in Table II, IL-2 was much less effective than T supt at supporting generation of ISC from SA-activated B cells. By contrast, substantial numbers of ISC were generated in response to IL-2 when B cells were initially activated by SA + T supt. Similar findings were obtained when highly purified FACS selected B cells were utilized. In additional experiments not shown, B cell DNA synthesis supported by IL-2 was greatly augmented when B cells had been activated by SA in the presence of T cell supt.

The effect of lymphokines during the initial activation with SA. The aforementioned results suggested that the presence of T supt during the initial activation induced subsequent B cell responsiveness to IL-2. Since the T supt employed contained a number of activities including IL-2 and IFN-γ as well as B cell growth factor (BCGF) and differentiation factor (BCDF)-like activities, it was possible that one or another of these

Table II. The action of IL-2 in cultures of B cells
activated with SA or SA + T Supt.

B Cells	1st Incubation Stimulus	Addition during 2nd incubation		
		Nil	T Supt	IL-2
		(ISC per 10^3 B cells)		
Control	SA	0	182.4	6.6
	SA + T supt	1.4	251.2	147.2
FACS Selected	SA	0	104.0	12.6
	SA + T supt	2.4	209.6	136.0

1st Incubation: 48 hr. 2nd Incubation: 72 hr. r-IL-2 = 50 U/ml.

factors accounted for subsequent IL-2 responsiveness. Recombinant IL-2
and r-IFN-γ were therefore examined for their capacity to enhance responsiveness to IL-2 when present during the initial incubation with SA.
Again, it can be seen in Table III that inclusion of T supt during the
preincubation of B cells with SA resulted in a markedly increased subsequent generation of ISC in response to IL-2. The inclusion of either
r-IL-2 or r-IFN-γ instead of T supt during the initial activation with SA
also resulted in enhanced subsequent responsiveness to IL-2. In experiments not shown, both r-IL-2 and r-IFN-γ were also found to be effective
at preparing SA-activated B cells to proliferate in response to r-IL-2.
It should be pointed out that r-IFN-γ was unable to promote proliferation
or differentiation of SA-activated B cells irrespective of the presence
of T cell lymphokines during the initial activation (data not shown).

DISCUSSION

     These studies were carried out to examine the role of IL-2 in human
B cell responses. The data show that IL-2 could directly promote B cell
growth and differentiation into ISC indicating that both of these B cell
responses may be supported entirely by IL-2 as well as the more tradi-
tional B cell stimulatory lymphokines, BCDF and BCGF. The possibility
that IL-2 promoted the proliferation and differentiation of SA-activated
B cells as a result of an action of IL-2 on contaminating T cells was unlikely as similar responses were observed using B cells that were rigor-
ously depleted of T cells by negative selection with the FACS after exposure to OKT3 and OKT11. It is important to note that the methodology used
to prepare the B cells for these studies resulted in a markedly Mφ and NK
cell-depleted population. Removal of these potentially suppressive elements may be important in permitting the observation that IL-2 could support human B cell proliferation and differentiation maximally.

     A previously described two-step culture system (6) was used to deli-
neate the effects of IL-2 on B cell responses more precisely. In light of
the finding that IL-2 was just as effective as T supt in supporting both
growth and differentiation of SA-stimulated B cells when present from the
initiation of culture, it was somewhat surprising to find that IL-2 supported only minimal responses in cultures of B cells initially activated
by SA without T cell lymphokines. These results suggested that T cell
influences might play a necessary role during initial B cell activation
to permit subsequent IL-2 responsiveness, although responses supported by
crude T supt did not require such preparative T cell signalling. These
findings also indicated that the crude T supt contained lymphokines distinct from IL-2 (BCGF and/or BCDF?) that can promote responses of B cells
initially activated in the absence of T cell influences.

Table III. B cells initially activated with SA in the presence of r-IL-2 or r-IFN-γ are enhanced in subsequent responsiveness to r-IL-2.

Expt. No.	1st Incubation Stimulus	Addition during 2nd incubation		
		Nil	T supt	r-IL-2
		(ISC per 10^3 B cells)		
1	Nil	0	6.4	0
	SA	0	198.6	24.2
	SA + T supt	3.6	278.0	238.2
	SA + r-IL-2	3.4	206.8	181.2
	SA + r-IFN-γ	0.4	226.8	176.2
2	Nil	0	29.4	0.4
	SA	0	205.0	50.0
	SA + T supt	1.4	238.0	284.2
	SA + r-IL-2	0.8	260.4	187.2
	SA + r-IFN-γ	0	255.6	137.2

r-IL-2 and r-IFN-γ were used at final concentrations of 100 U/ml.

We had previously found that T supt could stimulate ISC generation from SA-activated post-divisional B cells, but an additional population of ISC precursors might require T cell-derived differentiative signals before initial division to secrete Ig subsequently in response to differentiation factors (6). The data described here indicate that the subset of B cells responding in this manner may have been the IL-2 responsive subpopulation since SA-activated B cells subsequently respond maximally to IL-2 only if T cell influences had been present during the initial activation. Importantly, the T cell signal(s) required to induce subsequent responsiveness to IL-2 could be provided by a conventional T cell supernatant, or by r-IL-2 itself, or r-IFN-γ. More recent studies suggest that BCGF may also be able to convey this initial B cell activation signal.

These data support the existence of two distinct pathways of human B cell activation. The first is one in which B cells activated by SA in the absence of T cell help respond maximally only to T supt but not IL-2. The second pathway is one in which B cells are activated by SA in the presence of a T cell-derived lymphokine such as IFN-γ or IL-2 that can deliver an early maturational signal. In this circumstance, subsequent proliferation and differentiation can be promoted by either IL-2 or T supt. IL-2 alone, therefore, can support both proliferation and differentiation of human B cells maximally when present from the initiation of the response. Thus IL-2 can play a major role in initiating and propagating the growth and differentiation of human peripheral blood B cells.

REFERENCES

1. Korsmeyer, S.K., W.C. Greene, J. Cossman, S. Hsu, J.P. Jensen, L.M. Neckers, S.L. Marshall, A. Bakhshi, J. Depper, W.K. Leonard, E.S. Jaffe, and T.A. Waldmann. 1983. Proc. Natl. Acad. Sci. USA. 80:4522.
2. Zubler, R.H., J.W. Lowenthal, F. Erard, N. Hashimoto, R. Devos, and H.R. MacDonald. 1984. J. Exp. Med. 160:1170.
3. Nakaniski, K., T.R. Malek, K.A. Smith, T. Hamaoka, E.M. Shevach, and W.E. Paul. 1984 J. Exp. Med. 160:1605.
4. Miedema, F., and C.J.M. Melief. 1985. Immunol. Today 6:258.
5. Thiele, D.L., and P.E. Lipsky. 1983. J. Immunol. 131:2282.
6. Jelinek, D.F., and P.E. Lipsky. 1985. J. Immunol. 134:1690.

APPLICATION OF THE STAPHYLOCOCCUS AUREUS BACTERIA TO THE STUDY

OF HUMAN B CELL PROLIFERATION AND DIFFERENTIATION

Sergio Romagnani

Division of Allergology and Clinical Immunology
University of Florence
Policlinico di Careggi, 50134 Firenze, Italia

INTRODUCTION

The ability of several bacteria or bacterial cell products to act as polyclonal B-cell activators (PBAs) is known for many years. Among bacterial PBAs active on human B cells, Staphylococcus aureus Cowan first strain (SAC) and its cell wall component, Staphylococcal protein A (SpA), is undoubtedly the most interesting one. In the first part of of this paper the mechanisms by which SAC bacteria activate human B cells are discussed. In the second part, data obtained by the application of SAC bacteria to the study of events involved in the induction of human B cells from the resting state to immunoglobulin (Ig)-secreting cells are reported.

MECHANISMS OF HUMAN B CELL ACTIVATION BY SAC BACTERIA

In 1976 Forsgren et al.[1] first showed that SAC bacteria were capable of inducing significant proliferation by human lymphocytes. Owing to the well known property of SpA to interact with the Fc region of human IgG, it was obvious to suggest, as these authors did, that the mitogenicity of SAC bacteria was related to the interaction of SpA present on the bacterial cell wall with the Fc region of IgG present on the surface of human B lymphocytes.

The mitogenicity of SAC bacteria is due to SpA present on the bacterial cell wall.

In 1978 we demonstrated that both SAC bacteria and SpA linked to Sepharose beads promoted significant proliferation by unfractionated tonsil lymphocytes and their mitogenic activity was even higher on purified B cells. In contrast, soluble SpA showed detectable mitogenic activity only on unfractionated tonsil cell populations. Readdition of increasing concentrations of autologous T cells to purified B cells did not induce any further increase of the SAC-induced proliferative response, whereas it restored the B cell proliferation in the presence of soluble SpA[2]. Based on these data, we concluded that SpA linked to an insoluble matrix, such as SAC bacteria or Sepharose beads, acted as a T-cell independent B-cell

67

mitogen, whereas soluble SpA needed the presence of both T and B cells to express its mitogenic activity. Similar results were obtained when peripheral blood instead of tonsil lymhocytes were used.

However, in considering the mechanism responsible for B-cell activation by insoluble SpA as related to an interaction between SpA and the Fc region of surface IgG, two findings were disturbing. First, the number of human B cells equipped with surface IgG in peripheral blood was too much small (1% or less) to justify the strong proliferation induced by SAC bacteria on day 3. Second, we observed that a large proportion of both blood and tonsil human B cells (between 30 and 50%) was capable of binding directly fluoresceinated bacteria. This number was clearly greater that that of surface IgG-bearing cells possibly present in peripheral blood or tonsil cell suspensions. We attempted, therefore, to clarify this apparent paradox by investigating better the nature of the surface component(s) responsible for the reactivity of a proportion of human B cells with SpA present on SAC bacteria. This was done by an extensive series of experiments carried out from 1979 to 1982.

SpA binds not only to B-cell surface and serum IgG, but also to a proportion of serum IgM and to surface IgM from a subset of human B cells

We first investigated the phenotype of human B cells capable of reacting with SpA by using a rosetting technique with SpA-coated erythrocytes. We found that both IgG-bearing, as well as a considerable proportion of IgM-bearing, normal human B cells were able to react with insolubilized SpA[3]. In addition, we studied the SpA-binding ability of monoclonal B cells from patients with chronic lymphocytic leukemia (CLL). Virtually all B cells from 15 of 38 patients examined showed the ability to bind SpA, whereas B lymphocytes from 23 patients did not. In 14 patients the SpA-reactive B cells bore surface IgM or IgM and IgD, whereas in one patient they apparently expressed surface IgG only [4]. We then demonstrated that incubation of normal purified B cells from tonsil with $F(ab')_2$ fragments of anti-$F(ab')_2$ antibodies virtually abolished the ability of these cells to form rosettes with SpA-coated erythrocytes, whereas the incubation with $F(ab')_2$ fragments of anti-μ or anti-γ antibodies resulted in a significant reduction of the number of SpA rosettes. In contrast, the incubation of cells with $F(ab')_2$ fragments of antibodies against human β-2 microglobulin had no effect [3]. This type of experiments was then repeated using SpA-binding monoclonal B cells from two patients with CLL. One of these patients had leukemic B cells bearing surface IgM and IgD with κ light chain, whereas lymphocytes from the other patient had surface IgG with λ light chain. The incubation with anti-γ antibodies abolished the SpA reactivity by surface IgG-bearing leukemic cells only, whereas the incubation with anti-μ antibodies had no effect on the SpA rosetting of surface IgG-bearing lymphocytes, but resulted in a significant decrease of the number of SpA rosettes formed by cells of the patient whose peripheral blood lymphocytes bore surface IgM and IgD [4]. These results suggested that SpA was able to bind not only to surface IgG, but also to surface IgM.

To further substantiate this point, the SpA-binding capacity of monoclonal IgM proteins, isolated from the serum of 13 patients with Waldenström macroglobulinemia was examined. Four of these macroglobulins bound remarkable amounts of radiolabelled SpA, whereas the other 9 did not. Interestingly, the proportion of SpA-reactive serum macroglobulins (31%) was equivalent to the proportion (35%) of CLL patients whose surface IgM-bearing leukemic B cells displayed the ability to form rosettes with SpA-coated erythrocytes.

SpA interacts with human Igs in two ways: classical interaction with Fcγ and alternative F(ab')₂ interaction with at least four Ig classes

The site of IgM proteins responsible for the reactivity with SpA was then investigated. IgM monomers, as well as F(ab')$_2$ fragments of two monoclonal IgM proteins still displayed the ability to bind remarkable amounts of radiolabelled SpA, even though to a lower extent than IgM pentamers. The ability of Fcμ fragments to bind SpA could not be evaluated [5]. To establish whether the IgM-reactive site on SpA was the same responsible for the SpA-Fcγ reactivity, we made use of a binding-inhibition assay based on the ability of IgG molecules from different species or their fragments to inhibit the reactivity of insolubilized SpA with radiolabelled human IgG, rabbit IgG or human IgM monomers. The binding of human IgG to SpA was completely inhibited by human IgG and partially by rabbit IgG and human Fcγ , but it was poorly or not inhibited by IgM, F(ab')$_2$μ or F(ab')$_2$γ fragments. Conversely, the binding of rabbit IgG to SpA was inhibited by human IgG, human Fcγ or rabbit IgG, but not by IgM, F(ab')$_2$μ or F(ab')$_2$γ. In contrast, the SpA-IgM binding was not inhibited by rabbit IgG or human Fcγ , but it was significantly inhibited or abolished by IgM, human IgG, F(ab')$_2$μ and F(ab')$_2$γ fragments. Similar data were also reported by Inganas et al. [6,7], supporting the concept of two distinct Ig-binding regions on SpA. The first is represented by the classical Fcγ-binding site; the other is capable of reacting with a proportion of human IgG, IgM, IgA and IgE via the interaction with a structure apparently located in the F(ab')$_2$ part of these Igs. The latter has been called the alternative Ig-binding site of SpA [6,7].

The mitogenicity of SAC bacteria on human B cells is primarily due to the interaction between the alternative binding site of SpA and surface IgM

In other experiments we attempted to ascertain whether SpA present on the bacterial cell wall was responsible for the mitogenic activity of SAC bacteria on human B cells and to establish the respective role of the two Ig-binding sites of SpA on the mitogenicity of SpA-containing SAC bacteria. We first demonstrated that B cells which proliferate in response to SAC bacteria were contained in the pool of SpA-binding cells. Part of them bore surface IgG, but the majority of these cells were equipped with surface IgM and IgD [8]. Then, we compared the mitogenic activity on human B cells of affinity-purified F(ab')$_2$ fragments of anti-μ antibodies and of SAC bacteria. The response of purified B cells to anti-μ and SAC bacteria was similar and increased in the presence of both reagents when cell suspensions were enriched in surface IgM-bearing B lymphocytes. In addition, monovalent fragments of anti-μ antibodies were able to inhibit both the anti-μ- and, even to a lower degree, the SAC-induced B cell proliferation [5]. Finally, we tried to establish whether the IgM-binding site of SpA was primarily responsible for the SAC-induced B cell proliferation. This was done by using two different experimental approaches. In the first, we studied the inhibitory effect on SAC-induced B cell proliferation of molecules showing reactivity of the Fcγ type, such as rabbit IgG, reactivity of the alternative type, such as SpA-reactive monoclonal IgM or both Fcγ- and F(ab')$_2$-reactivity, such as human IgG. Either polyclonal human IgG or SpA-reactive monoclonal human IgM inhibited the SAC-induced B cell proliferation, whereas polyclonal rabbit IgG molecules were poorly or not at all inhibitory [9].

In another series of experiments, we tried to inhibit selectively the Fcγ-binding region of SpA present on the bacterial cell wall by inactivating the thyrosil residues of SpA by a hyperiodination procedure. This treatment had been shown capable of abolishing the Fcγ-type reactivity of both soluble SpA and SAC bacteria without affecting the reactivity of the alternative type. We found that the inactivation of thyrosil residues of SpA had no effect on the mitogenic activity of both SAC bacteria and SpA coupled to Sepharose beads on human B cells [9].

Thus from the complex of these data, we concluded that SAC bacteria act as selective T-cell independent B cell mitogen via an interaction between the alternative Ig-binding site of SpA present on the bacterial cell wall and IgM present on the surface from a subset of human B cells.

APPLICATION OF SAC BACTERIA TO THE STUDY OF EVENTS INVOLVED IN PROLIFERATION AND DIFFERENTIATION OF HUMAN B CELLS

The ability of SAC bacteria to act on human B cells in a manner similar to that of insolubilized anti-IgM antibodies has been widely utilized to examine the sequence of events involved in the induction of human B cells from the resting state to Ig-secreting cells. Two assay systems just based on the use of anti-μ antibody and SAC bacteria were established. The first system used SAC stimulation for 2-3 days, followed by incubation of B cells with T cell-derived factors for additional 3 days [10]. The second system was a 3-day co-stimulation assay with anti-μ antibody and T cell factors, analogous to the co-stimulation assay already employed in mice [11]. On the basis of studies performed by the two assay systems, a model for human B cell proliferation and differentiation was proposed. In this model, the initial signal is provided by the cross-linking of surface Ig, such as that delivered by anti-μ , SAC or antigen. This initiates B cell activation and the cell expresses receptors for growth factors. In the presence of the appropriate growth factors, the cell enters S phase and becomes a cycling cell. Subsequently, receptors for differentiation factors are expressed and the presence of those factors initiates the production and secretion of Igs [12]. In the last two years the nature and the role of some T cell-derived factors supporting human B cell responses have been extensively investigated in our laboratory.

Anti-IgM-activated human B cells proliferate in response to both B cell growth factor (BCGF) and interleukin 2 (IL-2), whereas SAC-activated B cells respond to IL-2, but not to BCGF

In a first series of experiments, we compared the reactivity of human B cells to IL-2 in the co-stimulation assay with anti-μ antibody and in the pre-activation assay with SAC bacteria. To this end, we used a purified IL-2, obtained by the recombinant DNA technology (r IL-2; Biogen,

Geneve). In agreement with recently reported data [13-15], we found that both anti-μ- and SAC-activated B cells exhibited substantial proliferation in response to r IL-2. An urgent question was, therefore, to re-examine the role of previously described BCGF(s). To this end, an attempt was made to establish human T cell clones capable of producing BCGF, but not IL-2. This was done by applying the limiting dilution microculture conditions developed by Moretta et al [16]. T cell clones so obtained were stimulated with phytohemagglutinin (PHA) or SpA and their supernatants examined for both BCGF and IL-2 activity. Supernatants from more than a hundred clones were screened: the great majority of those from T4[+], but also a large proportion from T8[+] clones, were found to display both BCGF and IL-2 activity. Supernatants from some clones, however, manifested BCGF activity in the co-stimulation assay with anti-μ antibody, apparently in the absence of IL-2 [17].

To provide additional evidence that BCGF activity present in supernatants from these clones was related to growth factors distinct from IL-2, we investigated the effect of anti-TAC antibody on the proliferative response of anti-μ-activated B cells cultured in the presence of the supernatant from one of these clones. As expected, the anti-TAC monoclonal antibody remarkably inhibited in a dose dependent fashion the r IL-2-induced B cell proliferation, but had little or no effect on the B cell proliferation promoted by the IL-2-free, BCGF-containing, clonal supernatant. The BCGF activity of 3 clonal supernatants lacking IL-2 activity was then paralleled assessed by the co-stimulation assay with anti-μ antibody and the pre-activation assay with SAC bacteria. Unexpectedly, supernatants from all 3 clones, which displayed substantial BCGF activity when assayed in the co-stimulation system with anti-μ antibody, manifested little, if any, potentiating effect on the proliferative response of SAC-activated B cells [17].

BCGF and IL-2 act on different phases of human B cell activation

The above data suggest that activated human B cells react in different manner to IL-2 and BCGF in the two assays commonly used to assess the BCGF activity. While the co-stimulation assay with anti-μ antibody explores the reactivity of B cells to both IL-2 and BCGF(s) distinct from IL-2, the pre-activation assay with SAC bacteria seems to represent mainly a probe for the study of B cell reactivity to IL-2. The next question addressed was, therefore, to establish whether this was due to the different nature of activating signals provided by anti-μ antibody and SAC or reflected differences in the time of lymphokine addition to the culture between the two assays. To solve this problem, purified B cells were stimulated for 3 days with anti-μ antibody and this was followed, like in the SAC assay, by incubation with r IL-2 or BCGF-containing supernatants for additional 3 days. As expected, the addition of both r IL-2 and BCGF potentiated the proliferative response of human B cells in the 3-day co-stimulation assay with anti-μ antibody. Furthermore, r IL-2 showed the ability to promote proliferation of B cells pre-activated for 3 days with either SAC bacteria or anti-μ antibody, whereas BCGF virtually lost its effect when added to anti-μ- or SAC-preactivated B cells 3 days after the initiation of culture [17]. These results clearly demonstrate that the different reactivity of activated human B cells to BCGF and IL-2 in the two assay systems was not due to differences in the quality of activation signals (anti-μ or SAC). More likely, it means that following initial activation, B cells maintain their reactivity to BCGF for a shorter time than to IL-2. This possibility was further substantiated by other experiments in which we demonstrated that B cells activated for 24 hr with anti-μ antibody were more effective than B cells activated for 72 hr with SAC bacteria in adsorbing the BCGF activity from a clonal IL-2-free, but BCGF-containing, supernatant (unpublished data). Although indirectly, this suggests that the hypothetical receptor for BCGF is expressed on a larger number of B cells or is present at higher density within the first 24 hr since initial activation.

In subsequent experiments, the effect of r IL-2 and the clonal BCGF preparation added simultaneously to anti-μ-stimulated B cells was evaluated. When r IL-2 and BCGF were added together at the beginning of culture the B cell response was close to the sum of responses induced by the two factors added separately. Such a potentiation was not observed using concentrations of IL-2 even four times greater than those used in the combined experiment [18]. When the response of anti-μ-activated B cells to r IL-2 and clonal supernatant with BCGF activity added sequentially was investigated. A synergistic response was found if BCGF was added at the beginning and IL-2 after 24 or 48 hr of culture. In contrast, when IL-2 was added at the beginning and the addition of BCGF was delayed, no synergy in the activity of the two factors was observed [18]. These results provide additional evidence that BCGF receptors are expressed on activated B cells for a shorter time period than receptors for IL-2 and suggest that the proliferation of B cells is determined by sequential interaction of BCGF and IL-2 with their respective receptors.

Gamma interferon (IFN-γ) can act as a growth factor for anti-μ, but not for SAC-, activated human B cells

To further investigate the role of molecules produced by T cells on B cell proliferation, in a subsequent series of experiments we examined the effect on the proliferative response of anti-μ- or SAC-activated B cells of recombinant IFN-γ (r IFN-γ; Biogen, Geneve). These experiments have been performed in collaboration with M.C. Mingari, L. Moretta and C.M. Liang. Highly purified B cells stimulated with anti-μ antibody could be induced to proliferate with r IFN-γ. In contrast, when r IFN-γ was tested for BCGF activity in the pre-activation assay with SAC bacteria, no potentiating effect on the SAC-induced B cell proliferation was observed (Fig. 1). This finding parallels that obtained with clonal supernatants containing BCGF, but not IL-2, and provides additional evidence that when IL-2 receptors are widely expressed on activated B cells, these latter lose the ability to respond to BCGF(s) distinct from IL-2. When r IFN-γ and IL-2 were combined, the total proliferative response of anti-μ-activated B cells was further increased, thus suggesting a synergy between IFN-γ and IL-2 in promoting proliferation of anti-μ-activated B cells.

The evidence that r IFN-γ induced B cells to proliferate and, like some clonal BCGF preparations, had a synergystic effect in the IL-2-promoted B cell proliferation, prompted us to ask whether the BCGF activity

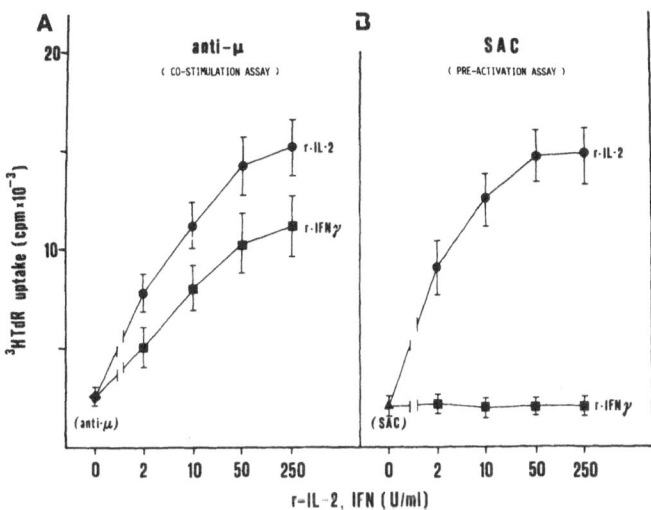

Fig. 1. r IFN-γ promotes B cell proliferation by anti-μ-, but not by SAC-, activated human B cells. (a) Highly purified tonsillar B cells (1 x 10⁵/well) were stimulated for 3 days with suboptimal concentrations of anti-μ antibody (5 μg/ml) in the presence of different concentrations of r IL-2 or r IFN-γ. (b) B cells (1 x 10⁵/well) were stimulated for 3 days with SAC bacteria and cultured for additional 3 days with different concentrations of r IL-2 or r IFN-γ. Sixteen hr before harvesting, B cell cultures were pulsed with 0.5 μCi of ³H-thymidine (³H-TdR).

Table 1. Inhibitory effect of anti-IFN-γ and anti-TAC monoclonal antibodies on the BCGF activity of a crude polyclonal T cell conditioned supernatant[a]

Stimulant	Monoclonal antibody added			
	Control ascites	Anti-IFN	Anti-TAC	Anti-IFN + anti-TAC
	cpm	cpm	cpm	cpm
r IFN-γ	14,274	4,800	n.d.	4,900
r IL-2	16,358	16,027	5,200	5,100
Crude SUP	15,747	13,052	10,811	6,838

[a]Purified B cells were stimulated for 3 days with anti-μ antibody (5 μg/ml) and r IFN-γ (50 U/ml), r IL-2 (25 U/ml) or crude supernatant (1:8), in the absence or in the presence of monoclonal anti-IFN-γ antibody (1:1,000), anti-TAC antibody (1:500) or both. The mean value of cpm in cultures stimulated with anti-μ alone was 4,974. n.d. = not determined.

of crude polyclonal T cell conditioned supernatants was due, at least in part, to IFN-γ. To this end, we employed 3 different monoclonal anti-bodies and a rabbit polyclonal antibody specific for IFN-γ to neutralize the IFN-γ activity in these supernatants. The anti-TAC monoclonal anti-body was used to inhibit the activity of IL-2. The results of a repre-sentative experiment are reported in Table 1. While incubation with anti-IFN-γ had slight inhibitory effect and incubation with anti-TAC alone partially inhibited the BCGF activity of the crude polyclonal T cell supernatants, the simultaneous neutralization of IFN-γ and IL-2 resulted in a marked, even if not absolute, inhibition of the BCGF activity of the crude BCGF preparation. Similar results were obtained with 3 different crude preapartions, thus suggesting that IFN-γ and IL-2 together account for most of the BCGF activity contained in conventional polyclonal T cell derived supernatants.

Obviously, these data do not exclude the existence of BCGF(s) dis-tinct from IFN-γ or IL-2. BCGF(s) showing physical and chemical features different from those of both IFN-γ and IL-2 have been demonstrated in supernatants from T cell hybridomas and T cell clones [19,20]. In addition, we recently found that the supernatant from a T cell clone (DP 5/11), containing BCGF but not IL-2, maintained most of its BCGF activity after removal of IFN-γ activity (manuscript in preparation).

Taken together, our results indicate that at least 3 molecules with BCGF activity do exist: IL-2, IFN-γ and another factor distinct from IL-2 and IFN-γ, that we can continue to call BCGF until it is purified to biochemical homogeneity or its gene is cloned. The so-called BCGF and IFN-γ seem to be involved in the initial phases of B cell activation and their activity is easily demonstrable by the co-stimulation assay with anti-μ antibody. These factors probably act by favouring or enhancing the B cell reactivity to IL-2. For example, they can induce the expres-sion or increase the density of IL-2 receptors on activated B cells [18]. In contrast, as is well shown by the pre-activation assay with SAC bacte-ria, the activity of IL-2 differs from that of BCGF and IFN-γ, since this mediator is mainly required for maintaining the continued proliferation of activated B cells.

IL-2 induces SAC-activated B cells not only to proliferate, but also to differentiate into Ig-secreting cells

While it is undoubt that SAC bacteria act as T cell independent mitogen for human B cells, their ability to induce human B cells to dif-ferentiate into Ig-secreting cells is conditioned by the presence in cul-ture of T cells or T cell factors [21,22].

To examine the role of IL-2 in the human B cell differentiation in-duced by SAC bacteria, we tested the ability of T cell- and monocyte-depleted populations from peripheral blood to produce IgM and IgG in vitro in the presence of r IL-2 alone or r IL-2 plus pokeweed mitogen (PWM) or SAC bacteria. As control, we also tested the activity on the same cell suspensions of PWM or SAC alone or added to the cultures together with various concentrations of autologous T lymphocytes. The results of these experiments are summarized in Fig. 2. Production of both IgM and IgG was never observed in cultures containing PWM, SAC bacteria or r IL-2 alone.

B cells activated with SAC bacteria produced considerable amounts of Igs when a concentration of r IL-2 as low as 1 U/ml was added. In contrast, the addition of r IL-2 to cultures stimulated with PWM (a T cell-dependent B cell activator) gave poor or no production of Igs. The addition of as few as 5-10% autologous T cells restored the PWM-induced Ig production by B cells, whereas it had poor or no effect on the Ig production of cultures stimulated with SAC bacteria in the absence of IL-2. These results clearly indicate that the interaction with IL-2 is sufficient to induce significant production of IgM and IgG, provided that an efficient signal, such as that delivered to human B cells by SAC bacteria, is utilized.

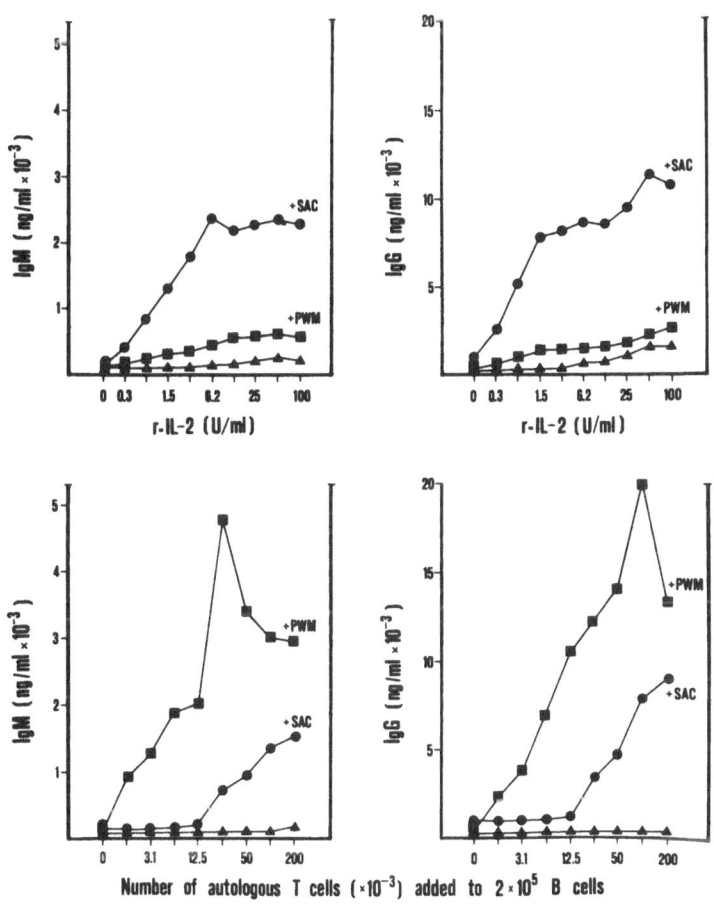

Fig. 2. IL-2 induces SAC-, but not PWM-, activated B cells to differentiate into Ig-secreting cells. B cell-enriched peripheral blood suspensions (2 x 10^5/tube) were cultured for 7 days in the presence of optimal concentrations of SAC or PWM, without or with different concentrations of r IL-2 (top) or autologous T lymphocytes (bottom). IgM and IgG released in the culture supernatants were measured by appropriate radioimmunoassays, previously described in detail [23].

To establish whether IL-2 was also able to induce in vivo activated human B cells to differentiate into Ig-secreting cells, highly purified tonsillar B cells were separated on the basis of their density by centrifugation on Percoll density gradients. The higher density cell fraction mainly consisted of small B cells, whereas the lower density population was made up of somewhat larger cells, most of which are known to be activated B cells. Unfractionated B cells, as well as their high and low density fractions were then tested for ability to produce IgM and IgG in vitro following stimulation with r IL-2 alone. Unfractionated B cells from tonsil of several donors produced detectable amounts of both IgM and IgG when incubated for 7 days with r IL-2. The Ig production became detectable after 4 days of culture and the peak production was usually shown to occur on day 7. R IL-2 induced Ig production by purified tonsil B cells in a dose-dependent fashion and its activity was inhibited by addition to the cultures of anti-TAC antibody. Interstingly, production of both IgM and IgG was reduced in cultures mainly consisting of high density small B cells, whereas it was markedly enhanced in those containing low density,larger, in vivo activated, B cells [24].

Thus, from the complex of these data, we conclude that IL-2 plays a direct role also in the differentiation of human B lymphocytes into antibody-producing cells.

CONCLUDING REMARKS AND SUMMARY

In this paper we have summarized our work on the mechanisms by which SAC bacteria act as a T cell independent B cell mitogen for human B cells and on the application of SAC bacteria to the study of events involved in the induction of resting B cells to Ig-secreting cells. We have shown that the mitogenic activity of SAC bacteria on human B cells was due to an interaction between the alternative Ig-binding site of protein A present on the bacterial cell wall and surface IgM from a subset of B lymphocytes. Cross-linking of surface IgM by SAC bacteria resulted in a strong, but time-limited, proliferation of the reactive B cells. The proliferative response induced by SAC bacteria could be maintained by addition to B cell cultures of IL-2-containing T cell conditoned supernatants or recombinant IL-2. Additionally, a small concentration of IL-2 directly induced in vitro SAC-stimulated or in vivo activated B lymphocytes to differentiate into Ig-secreting cells. In contrast, BCGF(s) distinct from IL-2 and recombinant IFN-γ promoted the proliferative response of B cells activated with suboptimal concentrations of anti-μ antibody, but were unable to maintain the B cell proliferation induced by SAC bacteria. However, both recombinant IFN-γ and another BCGF distinct from IFN-γ and IL-2, if added to B cell cultures together with, or before than, IL-2, had a potentiating effect on the IL-2-promoted proliferation of anti-μ-activated B cells.

These data suggest that the early interaction of activated B cells with BCGF(s) distinct from IL-2 (including IFN-γ) positively affects the response of these cells to IL-2, which seems to be the mediator primarily implicated in maintaining the continued proliferation of the activated B cells, as well as their differentiation into Ig-producing cells.

ACKNOWLEDGMENTS

The experiments reported in this paper were supported by grants from CNR (Finalized Projects 'Infectious Diseases' and'Oncology') and from AIRC. I thank my co-workers Drs. A. Alessi, F. Almerigogna, R. Biagiotti, G.F. Del Prete, M.G. Giudizi, M. Mazzetti, D. Macchia, E. Maggi, P. Parronchi, A. Tiri and D. Vercelli, whose contributions have been essential in the experiments reported in this paper.

REFERENCES

1. Forsgren, A., Svedjelund, A. & Wigzell, H. Eur. J. Immunol. 6, 207-212 (1976)

2. Romagnani, S., Amadori, A., Giudizi, M.G., Biagiotti, R., Maggi, E. & Ricci, M. Immunology 35, 471-477 (1978)

3. Romagnani, S., Giudizi, M.G., Almerigogna, F. & Ricci, M. J. Immunol. 124, 1620-1626 (1980)

4. Romagnani, S., Giudizi, M.G., Biagiotti, R., Almerigogna, F., Del Prete, G.F., Maggi, E. & Ricci, M. Scand. J. Immunol. 15, 287-295 (1982)

5. Romagnani, S., Giudizi, M.G., Biagiotti, R., Almerigogna, F., Maggi, E., Del Prete, G.F. & Ricci, M. J. Immunol. 127, 1307-1313 (1981)

6. Inganäs, M. Scand. J. Immunol. 13, 343-352 (1981)

7. Inganäs, M. & Johansson, S.G.O. Int. Arch. Allergy appl. Immunol. 65, 91-101 (1981)

8. Romagnani, S., Giudizi, M.G., Almerigogna, F., Nicoletti, P.L. & Ricci, M. Immunology 39, 417-425 (1980)

9. Romagnani, S., Giudizi, M.G., Del Prete, G.F., Maggi, E., Biagiotti R., Almerigogna F. & Ricci, M. J. Immunol. 129, 596-602 (1982)

10. Muraguchi, A. & Fauci, A.S. J. Immunol. 129, 1104-1108 (1982)

11. Howard, M., Nahanishi, K. & Paul, W.E. Immunol. Rev. 78, 185-205 (1984)

12. Kehrl, J.H., Muraguchi, A., Butler, J.L., Falkoff, R.J.M. & Fauci A.S. Immunol. Rev. 78, 75-96 (1984)

13. Tsudo, M., Uchiyama, T. & Uchino, H. J. exp. Med. 160, 612-617 (1984)

14. Mingari, M.C., Gerosa, F., Carra, G., Accolla, R.S., Moretta, A., Zubler, R.H., Waldmann, T.A. & Moretta, L. Nature (London) 312, 641-643 (1984)

15. Muraguchi, A., Kehrl, J.H., Longo, D.L., Volkman, D.J., Smith, K.A. & Fauci, A.S. J. exp. Med. 161, 181-197 (1985)

16. Moretta, A., Pantaleo, G., Moretta, L., Cerottini, J.L. & Mingari, M.C. J. exp. Med. 157, 743-747 (1983)

17. Almerigogna, F., Biagiotti, R., Giudizi, M.G., Del Prete, G.F., Maggi, E., Mazzetti, M., Alessi, A., Ricci, M. & Romagnani, S. Cell. Immunol. 95 (1985, in press)

18. Romagnani, S., Giudizi, M.G., Maggi, E., Almerigogna, F., Biagiotti, R., Del Prete, G.F., Mazzetti, M., Alessi, A., Vercelli, D. & Ricci, M. Eur. J. Immunol. (1986, in press)

19. Butler, J.L., Ambrus, J.L. & Fauci, A.S. J. Immunol. 133, 251-255 (1984)

20. Yoshizaki, K., Nakagawa, T., Fukunaga, K., Kaieda, T., Maruyama, S., Kishimoto, S., Yamamura, Y. & Kishimoto, T. J. Immunol. 130, 1241-1246 (1983)

21. Falkoff, R.J.M., Zhu, L.P. & Fauci, A.S. J. Immunol. 129, 97-102 (1982)

22. Saiki, O. & Ralph, P. J. Immunol. 127, 1044-1049 (1981)

23. Romagnani, S., Del Prete, G.F., Maggi, E., Bellesi, G., Biti, G., Rossi-Ferrini, P.L. & Ricci, M. J. Clin. Invest. 71, 1375-1382 (1983)

24. Romagnani, S., Del Prete G.F., Giudizi, M.G., Biagiotti, R., Almerigogna, F., Tiri, A., Alessi, A., Mazzetti, M. & Ricci, M. (submitted for publication)

A MODEL FOR THE FIRST ACTIVATION CYCLE OF HUMAN B LYMPHOCYTES

J. Gordon, G. Guy*, and L. Walker

Departments of Immunology and Biochemistry*
The Medical School
Birmingham, B15 2TJ, UK

The human B lymphocyte provides an excellent model for examining cell cycle control and its undermining by malignant transformation. Resting B cells can be isolated in both high yield and purity and their response to a variety of defined stimuli followed in detail. Of particular interest is the activator Epstein-Barr virus (EBV) which infects human B cells via their receptors for the C3d fragment of complement (CR2) and subsequently bestows immortality through expression of the viral genome[1]. P3HR-1, a mutant strain of EBV, while retaining CR2-binding capacity, lacks a deletion in the EBNA2 transforming region and fails to induce growth in infected cells[2]. In this report we detail the outcome of exposing resting B cells to both the transforming and non-transforming strains of EBV within the context of the mitotic cell cycle and compare the results with those obtained from activators which use more physiological routes of B-cell triggering.

MATERIALS AND METHODS

EBV was the generous gift of Professor A Rickinson and used at a titred optimal concentration. SAC (Staphylococcus Aureus Cowan Strain I) was obtained from Calbiochem, Cambridge U.K., ionomycin from Calbiochem, La Jolla, CA and TPA from Sigma, Poole, U.K. Cell surface antigens were detected in an indirect rosetting system using the following antibodies: BK19.9, a gift of Dr G Brown; MHM6, a gift of Dr M Rowe; A2 to the transferrin receptor (Tf-R), a gift of Dr A Bernard; 11EF7, developed by Dr N Ling. G_0 B cells were prepared by extensive negative selections and a final collection of cells which banded through a 57.5% Percoll (Pharmacia, Uppsala, Sweden) density gradient. Such preparations were free of monocytes and T cells at a level of <0.2%. The labelling of cells with {^{3}H} inositol and the subsequent analysis of water soluble inositol phosphates were as described previously[3].

RESULTS AND DISCUSSION

Table 1 summarizes the data obtained on stimulating quiescent tonsillar B cells with the activators used in this study. Cross-linking of the B-cell antigen receptors with either anti-immunoglobulin or SAC resulted in a hydrolysis of inositol phospholipids leading to an accumulation of inositol trisphosphate (IP_3). Although at 5 min the levels of IP_3 were similar in B cells exposed to either activator, only SAC prompted cell cycle entry and its subsequent progression as revealed by RNA and DNA synthesis at appropriate time points.

Table 1. Sequels to the activation of G_0 B cells.

	cpm			% cells positive			
	IP_3 5 min	{³H}Urd 16-24h	{³H}Tdr 60-72h	BK19.9 6h	MHM6 6h	Tf-R 48h	11EF7 48h
Control	25	472	190	4	2	3	0
anti-Ig (15μgml⁻¹)	298	327	210	33	40	4	0
SAC (1:20,000)	365	2,935	135,392	42	46	45	18
EBV (B95-8)	27	967	98,738	52	54	40	35
EBV (P3HR-1)	nd	1,154	187	57	55	4	1
TPA (0.1ng ml⁻¹)	25	464	258	30	42	2	0
Ionomycin (0.8μg ml⁻¹)	26	483	85	40	42	5	0
TPA + Ionomycin	28	6,199	216,195	71	66	59	37

In addition to the accumulation of IP_3, which liberates intracellular calcium stores, degradation of inositol phospholipids generates diacylglycerol which activates the ubiquitous protein kinase-C. This bifurcating pathway provides a postulated "dual signal" for growth which can be examined with direct agonists of the individual arms[4]. At certain concentrations we noted that neither the phorbol ester TPA, which activates C-kinase, nor the calcium ionophore ionomycin was capable of getting resting B cells into cycle whereas together they not only provoked cell cycle entry but also induced a vigorous DNA response. The above observations imply a role for the "dual pathway" in the growth of human B cells and demonstrate that it can be activated through the perturbation of surface immunoglobulin. The results obtained with the transforming strain of EBV revealed, however, that involvement of inositol phospholipids was not obligatory for B-cell growth. Indeed, EBV induced high levels of DNA synthesis in B cells without any evidence of IP_3 accumulation or inositol phospholipid metabolism over the whole period of study (full data not shown). The P$_3$HR-1 virus deficient in the EBNA2 coding region, while failing to induce DNA synthesis managed to prompt G$_0$ B cells into early G$_1$ as evidenced by a small and transitory increase in RNA production.

All activators, including those which failed to get B cells into cycle, induced the expression of two antigens absent from the surface of quiescent cells. One of these is the B-cell specific Blast-2 antigen (recognized by MHM6) and the other is a ubiquitous "proliferation" antigen described by the monoclonal antibody BK19.9. The appearance of both antigens was remarkably rapid (Table 1). By contrast, the transferrin receptor (Tf-R) and a new B-cell activation antigen defined by 11EF7 appeared later and only on cells which had progressed through the cycle. In our hands, increased expression of class II (DR) antigen was a poor marker for early B-cell activation.

Figure 1 details the kinetics of antigen expression in G$_0$ B cells exposed to the full mitogenic signals of SAC and EBV. It is clear that during the first cell cycle, no phenotypic distinction can be made between cells which have received a full, but non-transforming, signal and those which are responding to a potentially oncogenic signal; nor is the outcome dependent on involvement of the inositol phospholipid pathway.

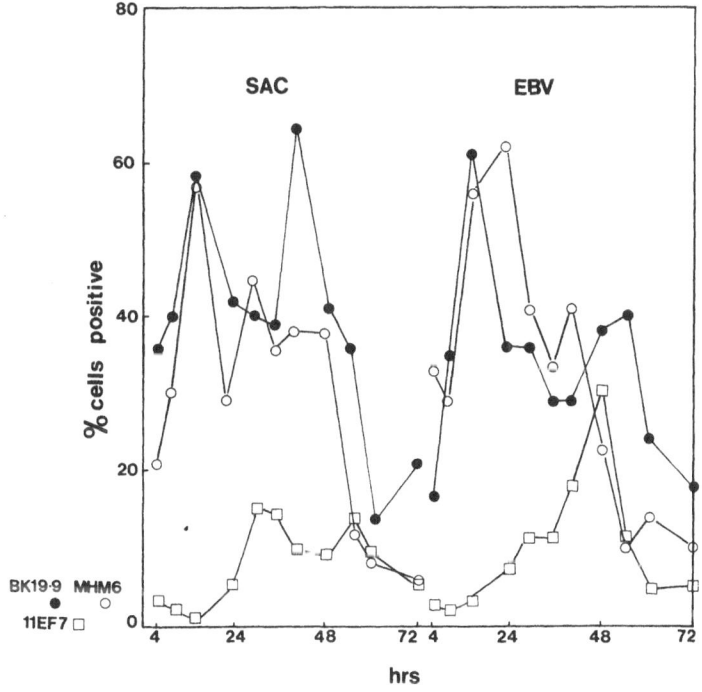

Figure 1. Kinetics of surface antigen induction on G$_0$ B cells exposed to EBV or SAC as detected in an indirect rosette assay. The values at 0 hr are shown to the left of the chart.

We are currently extending these studies by assessing cell cycle status through simultaneous analysis of RNA and DNA content in acridine orange stained cells. From the present accumulated data we can offer a model for the first activation cycle of B cells as shown in Figure 2. By detailing the genetic capacity of highly purified resting B cells to respond to a variety of defined stimuli we hope to provide a solid framework in which to investigate the role of growth and differentiation factors and explore the special relationship between human B cells and the immortalizing Epstein-Barr virus (see Fig. 2).

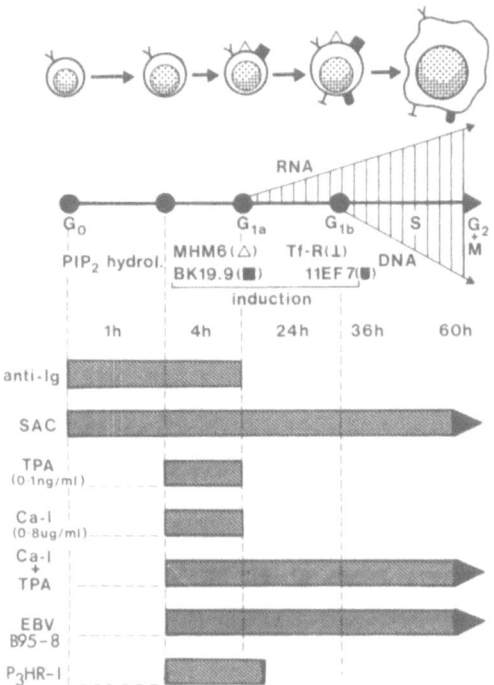

Figure 2. Cell cycle progression and associated phenotype of human G_0 B cells in response to
defined activators. PIP_2 hydrolysis indicates the breakdown in phosphatidylinositol
4,5-bisphosphate as measured by the accumulation of IP_3 (CaI = ionomycin).

ACKNOWLEDGEMENTS

This work was supported largely by grants from the MRC (UK).

REFERENCES

1. Bird, A.G., Britton, S., Ernberg, I. and Nilsson K.
    J. exp. Med. 154, 832 (1981)
2. Hennessy, K. and Kieff, E. Science 227, 1240 (1985)
3. Guy, G.R., Gordon, J., Michell, R.H. and Brown, G.
    Biochem. Biophys. Res. Commun. 131, 484 (1985)
4. Berridge, M.J. and Irvine, R.F. Nature 312, 315 (1984)

CONTROL OF B CELL ACTIVATION BY HELPER T CELLS

Kim Bottomly, John P. Tite, Barry Jones, Eileen Becker-Dunn, Jonathan Kaye, Abraham Kupfer, and Charles A. Janeway, Jr.

Division of Immunology, Department of Pathology
Howard Hughes Medical Institute at Yale University School of Medicine
New Haven, CT. 06510 USA

INTRODUCTION: STATEMENT OF THE PROBLEM

The activation of B cells to secrete antibody by helper T cells (Th) was one of the first cell interactions to be characterized in the immune response.[1] The process has been extensively characterized since its initial description twenty years ago. Two features of this cellular interaction have attracted particular attention, the requirement for physical linkage of the determinant recognized by the B cell to the determinant recognized by the helper T cell[2] and the requirement for identity in the I region of the major histocompatibility complex (MHC).[3] These two features suggested that intimate contact between the helper T cell and the B cell were required for efficient B cell activation. The stringency of these requirements, and the failure to activate 'bystander' B cells in in vivo systems, further suggested that a signal might be transmitted to the B cell directly via cell:cell contact, for instance by aggregation of the B cell's Ia glycoproteins.

In vivo analyses of the role of helper T cells in B cell activation have suggested, by contrast, that B cells may be activated in the absence of either linked recognition or I region matching. In such studies, the T cell must be activated by antigen presenting cells bearing the appropriate I region gene product and antigen, or by mitogenic lectins, but will then cause B cell activation presumably via the secretion of B cell activating factors.[4]

These two modes of activation, termed cognate and non-cognate, can both be observed in in vitro B cell responses. In particular, it has been noted that small resting B cells, and B cells from mice expressing the X-linked immuno-deficiency of the CBA/N mouse, require cognate interactions with helper T cells in order to become activated. It has been argued on this basis that these two pathways of activation are fundamentally distinct.[5] In this paper, we will argue that these two pathways of activation are fundamentally similar, and differ only in the efficiency of signalling from helper T cell to B cell. We will also describe different functional activities in subsets of helper T cells, all of which, however, appear to mediate their effects by the same basic mechanism.

In addition to helper T cells that interact with B cells via recognition of antigen:Ia complexes on the B cell surface, we noted in early studies that a second set of helper T cells, dependent upon B cells for their development, were required for optimal B cell activation in vivo.[6-11] In the antibody response to the hapten phospharylcholine (PC), which is normally dominated by the production of the idiotype associated with the BALB/c myeloma protein TEPC 15 (T15), we have shown that such cells synergize with conventional I region recognizing helper T cells to produce dominant T15 idiotype antibody responses.[11] Such apparently idiotype-specific Th have been termed ThId, in contrast to the MHC-restricted Th, termed ThMHC. Such ThId have been shown to bind idiotype on surfaces in a highly specific fashion[12], to interact with the B cell independent of the B cell's I region genotype, [13,14] and to require a second signal from antigen which need not be bound by the B cell [11-14] and which does not show conventional Ir gene control.[13] In this paper, we will examine different classes of idiotype-specific Th, including two cloned T cell lines that interact, by apparently different mechanisms, directly with the B cell's immunoglobulin receptor in the absence of self I region molecules and thereby alter the B cell's response.

RESULTS: OBSERVATIONS WITH CLONED HELPER T CELL LINES

Helper T Cell:B Cell Interactions are Cognate in In Vitro Antibody Responses to PC-Protein

We have used cloned, protein antigen specific, L3T4[+] T cell lines as a source of helper T cells. Their interactions with B cells have been analyzed in the antibody response to PC covalently linked to the appropriate protein carrier. The anti-PC antibody response to such PC-protein antigens in vivo is almost entirely comprised of antibodies bearing the T15 idiotypic determinant. T15-bearing anti-PC antibody is not produced in mice expressing the X-linked immunodeficiency of CBA/N mice, and for this reason, it is thought that T15-bearing anti-PC antibody is produced exclusively by those B cells capable of being activated in a non-cognate fashion. Despite this, we have observed that, when B cells of differing MHC genotype are mixed with PC-protein and a cloned Th that recognizes the protein in the context of the I region expressed by one of the B cells, all of the anti-PC antibody is produced by the B cells bearing the I region for which the cloned T cell line is specific.[15,16] Thus, by this criterion, such B cells can also be activated via cognate interactions, a result that has been confirmed by others using a different approach.[17]

It has been shown that cloned lines of helper T cells are sensitive to quantitative as well as qualitative differences in Ia glycoproteins on antigen presenting cells.[18-20] We have examined the role of Ia antigen density in T cell dependent B cell responses to PC as well. Using a monoclonal antibody that selectively depletes B cells of high Ia antigen density, it was shown that such B cells are preferentially activated by cloned Th.[16] Thus, by this even more stringent criterion, interactions between Th and B cells are cognate in this system.

Finally, the role of physical linkage of the hapten to the carrier has been examined in this system.[21] While high doses of carrier protein can give rise to polyclonal B cell activation (see below), efficient activation of PC-specific B cells under identical culture conditions occurs at antigen doses in which no polyclonal B cell activation occurs, and such responses require physical linkage of the hapten to the carrier. Thus, this in vitro antibody response closely resembles the earlier in vivo results, and confirms the cognate nature of the cellular interaction.

Ia Glycoproteins on the B Cell Surface are Important in Helper T Cell Activation, but do not Transduce a Signal to the B Cell

The results obtained in the antibody response to PC-protein generated by cloned helper T cell lines and normal B cells demonstrated that such cells discriminate between B cells based both on Th I region genotype of the B cell and the density of the Ia glycoproteins expressed on the B cell surface. This cognate cell interaction could imply that the helper T cell actually signals the B cell via the interaction of the T cell receptor with B cell surface Ia molecules.

Table 1.   Clone D10.G4.1 Preferentially Stimulates B Cells from Female Offspring of CBA/N (xid) Mothers Independent of the Mode of T:B Interaction

Responding B Cells	Stimulus	Proliferation (Δ CPM)		Ig Secretion (Δ PFC)	
		Female	Male	Female	Male
(CBA/NxBALB/c)F1	Conalbumin	110,000	95,000	42,000	7,000
(CBA/NxBALB.B)F1	I-A^b	82,000	61,000	46,000	11,000
(CBA/NxBALB/c)F1	3D3	177,000	92,000	56,000	5,200

4×10^5 purified B cells plus 2×10^4 mitomycin C-treated D10.G4.1 T cells were mixed with 100μg/ml conalbumin or 20ng/ml 3D3 monoclonal anti-T cell receptor antibody and cultured for 2 days for ^{3}H-Thymidine incorporation or 4 days for reverse plaque forming cells (PFC). Responses in the absence of antigen, 3D3, or for (CBA/NxBALB.B)F1 B cells, in the absence of D10 cells, subtracted to yieldΔ values.

To examine this question, we analyzed the polyclonal antibody responses generated by a cloned helper T cell line D10.G4.1, which could be stimulated to activate B cells in three distinct fashions: Recognition of the antigen conalbumin (CA) in the context of syngeneic I-A^k molecules, recognition of allogeneic I-A^b molecules, or by means of soluble anti-T cell receptor monoclonal antibody 3D3. We compared responses of small resting B cells with those of large, activated B cells, and also B cells expressing the xid gene.[22] All three stimuli led to equivalent B cell proliferative responses, and no difference was observed with the various B cell preparations (Table 1). However, when antibody production was measured, small B cells, and B cells from mice expressing the xid gene, gave about ten fold lower responses than those obtained with large B cells or B cells from normal mice (Table 1). Nevertheless, the important point to emerge from this study is that, whether the cloned T cell line interacted with the B cell via syngeneic or allogeneic Ia molecules, or independently of Ia recognition by means of an anti-T cell receptor monoclonal antibody, these differences occurred. Since the concentrations of anti-T cell receptor antibody used here rapidly modulate the expression of the T cell receptor,[23] it seems unlikely that the interaction of the T cell receptor with Ia molecules on the B cell transduces a significant signal to the B cell. It should be noted that antibodies to the Fc receptor on the B cell do not affect this mode of B cell activation. Thus, this study strongly suggests that Th signalling to B cells does not involve Ia molecules as signal transducers. It also reaffirms the importance of the nature of the B cell chosen for analysis on the results obtained in such studies.

Killing of Cloned B Lymphoma Cells by Cloned Th Cells Mimics B Cell Activation by Cloned Th Cells

Although the first demonstration of the requirement for self MHC recognition in cell interations involved B cell activation by helper T cells,[3] it was only when analogous results were obtained in assays of cytolytic T cells that such observations were widely accepted.[24] It seems likely that the major reason for this difference in perception is in the complexity of the helper assays, and in the time taken to observe the effect after mixing the histoincompatible T and B cells; by contrast, the cytotoxicity assay take three to four hours, and is widely believed to involve a simple T cell:target cell interaction free from immunoregulatory influences. Nevertheless, both approaches gave the same result, namely, that the T cell had to recognize antigen on the target cell surface in the context of self MHC glycoproteins.

The advantages of the cytotoxicity assay led to attempts to establish a similar assay for antigen-specific, Ia restricted helper T cells. It has been found that certain B cell lymphoma lines that express Ia antigens are efficiently killed by cloned T cell lines that can activate normal B cells to proliferate and secrete immunoglobulin. The specificity for antigen and self Ia recognition are identical in both systems.[25] The cytotoxicity assay has the advantage that it emplys two cloned populations interacting with one another, is rapid and readily quantitated, and allows one to recover and analyze those B cells that are resistant to the cytotoxic effect.

This assay has been used to demonstrate the following points. First, the cytotoxic effect requires the target cell to express the appropriate I region product as well as the protein antigen for which the T cell is specific. Second, there is a strict and stoichiometric relationship between Ia antigen density on the B cell target and susceptibility to killing by a given cloned line.[26] Thus, if one analyzes the target cell, one would argue that this is a cognate interaction, and it is strictly analogous to the results obtained in B cell activation studies. However, as was observed in polyclonal B cell activation, there is also killing of bystander target cells.[25,27] This, in turn, has allowed an examination of two important questions. First, is the expression of Ia glycoproteins on the target cells important for activation of the T cell to mediate cytotoxicity or for killing of the target, and second, is the mechanism of direct and of bystander killing the same or distinct?

The role of Ia glycoproteins has been shown to be critical in activation of the killer cell rather than in the killing process itself. Thus, Ia density correlates with the ability of a cell to activate the cytolytic process, but cells of varying Ia antigen density derived from a single parent B lymphoma are all equally susceptible to bystander cytolysis, or to cytolysis mediated by lectins.[26]

Evidence has also been developed that both direct and bystander cytolysis are mediated by the same process. This has been done by means of three different types of experiments. First, it has been shown that the extent of cytolytic activity directed at attached targets for an individual cloned T cell line is directly correlated with the ability of the same cloned T cell line to kill bystander target cells.[27] Second, a panel of targets was examined for its susceptibility to lectin-mediated cytolysis by the cloned cytolytic T cell lines. A range of susceptibilities was found, and this was directly related to the susceptibility of the same target cells to bystander cytolysis.[27] Third, we have immunoselected targets that are resistant to cytolysis but have normal levels of Ia antigens, and which induce normal or increased levels of bystander killing on other targets. These lysis-resistant targets can not be killed either by direct contact with cytolytic cloned T cell lines or as bystanders. However, these lysis resistant targets are fully susceptible to cytolysis mediated by class I specific cytolytic T cells and lymphokine activated killer cells (E. Grimm, personal communication).[28]

These studies are consistent with the hypothesis that at least some class II restricted cytolytic T cells kill sensitive targets by releasing lytic lymphokines, such as lymphotoxin[29] and the lytic lymphokine previously described by Quintans and Dick.[30] Indeed, all our cloned lines that are active in cytolysis release lymphotoxin, and all our cloned, class II restricted T cells that lack cytolytic potential do not release lymphotoxin.[27] Although cytolysis is not commonly though of as a function of class II MHC restricted T cells bearing L3T4, this activity is found in freshly isolated, antigen-stimulated lymph node T cells. Thus, the cytolytic activity observed in long term cloned T cell lines is not an artifact of prolonged culture. The in vivo relevance of this activity is now being tested in two experimental systems, an Ir gene system in which T cell proliferation is reciprocally correlated with antibody responsiveness and in which F1 hybrids show dominant T cell proliferative responses and recessive antibody responses (Tite and Foellmer, unpublished observations), and the transplantation of variant B lymphoma lines, in which Ia negative variants and lysis-resistant variants grow more rapidly than the parental tumor line (Khavari and Jones, unpublished observations).

Antigen-Specific Ia-Restricted Cloned T Cell Lines Suppress B Cell Responses in Linked Recognition

A third functional phenotype we have encountered among cloned, Ia-restricted, antigen-specific T cell lines is the ability to suppress B cell antibody responses. Cloned T cell lines of this type will proliferate in response to antigen:self Ia, will induce antigen dependent, Ia-restricted B cell proliferation, but fail to generate Ig secretion in either specific or polyclonal B cell responses. When combined with a cloned helper T cell line, such cells can suppress the plaque forming cell response under appropriate experimental conditions.[31] Similar results have been obtained by Asano and Hodes in a system that appears to be analogous, [32] as well as by Friedman et al.[33]

These cloned, Ia restricted T cell lines with suppressive activity do not resemble previously described suppressor T cells.[34] These cloned T cell lines are antigen-specific, Ia restricted and do not secrete long-range antigen binding suppressor factors. We have shown that such cells also can kill B lymphoma cells in an antigen-specific, Ia restricted fashion (West and Bottomly, unpublished observations). By taking advantage of the correlation of the T15 idiotype with B cells of low Ia antigen density, we have been able to show that such suppressor cells act selectively on B cells of high Ia density, and that they do so via linked antigenic recognition. Thus, the interaction of such cells with B cells precisely mimics the interaction of cloned helper T cells with B cells, but the functional outcome of the interaction is distinct.[31] We have not been able to determine whether the target of this suppression is the B cell itself, which could be killed by the suppressive cloned line, or an attached helper T cell, which could also be susceptible (Tite, unpublished observation).

Thus, this unusual suppressive Ia restricted cloned T cell line behaves in a fashion that is strictly analogous to cloned helper T cell lines; the distinction in functional outcome is most likely due to the release of a different set of lymphokines than are released by cloned helper T cell lines. We are presently investigating this possibility. There is no evidence that these cells differ in their receptor molecules or in their other cell surface markers from helper lines.

Cloned Helper T Cell Lines Can be Subdivided into Two Classes Differing in their Ability to Help T15-Bearing B Cells

We have examined the helper activity of a large number of cloned T cell lines in the antibody response to PC-proteins. The majority of these cloned T cell lines (Type 1) give rise to anti-PC antibody responses that have less that 50% T15+ antibody forming cells.[16,35] However, some cloned T cell lines will give rise to anti-PC antibody responses dominated by the T15 idiotype (Type 2). Limiting dilution analysis of B cell responses to PC-protein helped by these cloned Th lines demonstrates that the increase in T15 positive plaque forming cells is due to the activation of more T15 positive precursor B cells, as well as to increased burst size of such cells.[35]

Two possible mechanisms for this distinction are being examined. One possibility is that such cloned lines differ in their affinity for self Ia, the class II helper cells having a higher affinity and therefore more efficiently stimulating low Ia density B cells that are T15 positive. The alternative possibility is that these cloned lines differ in the lymphokines they release rather that in their receptors, and that the low Ia density B cells stimulate lower levels of lymphokine release and are therefore less likely to become activated to secrete antibody. We do not yet know the mechanism by which this difference in B cell activation occurs.

Helper T Cells Primed In Vivo in the Absence of B Cells Resemble the Majority of Cloned Helper T Cell Lines

When helper T cells are primed in vivo, two synergistically acting helper T cell types can be observed in several different experimental systems. For instance, in the adoptive secondary antibody response to the hapten, 2,4-dinitrophenyl (DNP), antibody responses are proportional to the second power of the number of helper T cells transferred.[6] However, if the helper T cells are derived from mice rendered B cell deficient by treatment with anti-u chain antibody from birth, then there is a linear correlation between antibody response and helper T cell number,[8,9] This finding suggested that a second class of helper T cells might exist, specific for immunoglobulin determinants on B cells, and this has proven to be the case in several different experimental systems.

The antibody response to PC-proteins, normally dominated by the T15 idiotype, comprises only 30% T15+ antibody forming cells if the helper T cells are derived from B cell deficient mice[11] or from mice genetically deficient in the expression response of MOPC 315 myeloma cells grown in diffusion chambers in vivo (Rohrer, Kemp, and Janeway, unpublished observations). This experimental maneuver allowed us to examine the nature of the cell apparently lost when T cells are primed in the absence of B cells.

When helper T cells from normal mice are mixed with helper T cells from B cell deficient mice, a synergistic effect is observed,[9,11] and the production of idiotype is restored. Thus, the lowered responses observed with Th from B cell deficient mice does not reflect the action of suppressor T cells. This has been confirmed by using anti-Lyt-2 treated cells in many experiments in a variety of systems. T cells from non-immune mice will not provide the necessary signal to augment idiotype expression. In examing the requirement for antigen priming, it was observed in all systems examined to date, that the second Th cell, which selectively augments the expression of idiotype, must be derived from an antigen-primed mouse and must be stimulated with antigen in order to mediate its effects.[8-14] This observation has now been repeated in the MOPC 315 tumor system as well. Because these helper T cells selectively augment idiotype expression, they have been termd ThId.

ThId, unlike conventional Th do not help B cells to secrete antibody but must collaborate with conventional, MHC restricted Th, or ThMHC. The interaction of ThId with B cells is not MHC restricted[14] and the priming with antigen is not under conventional Ir gene control.[13]

ThId Specifically Bind Idiotype, and Can Be Expanded In Vitro

As mentioned above, the majority of cloned, self-Ia recognizing helper T cell lines produced in our laboratories resemble Th primed in anti-μ treated mice in that they elicit anti-PC antibody responses that are not dominated by the T15 idiotype in vitro.[16,35] This in vitro system has allowed us to examine the specificity of ThId cells added back to these cultures. ThId were isolated from antigen-primed lymph node cells by passage over nylon wool columns, treatment with anti-Lyt-2 and adherence to T15-coated petri dishes.[12] The binding reaction could be inhibited by high concentrations of free T15, but not by even higher concentrations of a variety of monoclonal anti-PC antibodies that lacked expression of the T15 idiotype. Thus, such cells are clearly idiotype specific, and they can recognize idiotype in the absence of MHC antigens, consistent with their observed behavior in prior in vivo studies.

In order to further characterize such cells, they were expanded in vitro and cloned by limiting dilution. Clones were screened for the ability to augment T15+ anti-PC responses in cultures helped by a Type 1 ThMHC.[36] Several such cloned lines have now been analyzed. Because ThId appear to be quite different from ThMHC in several characteristics, the culture conditions developed for the efficient expansion of cloned ThMHC lines have not proved ideal for ThId growth. At present, we use IL-2 alone to expand these cells, which grow slowly under these conditions. T15 coupled to sepharose beads seems to selectively stimulate such cells, but strong proliferative or IL-2 responses to such T15 coated beads have not been observed. Antigen, if anything, inhibits ThId growth. Thus, we have not had the large numbers of ThId to analyze that one would like, but we have been able to characterize the biology of such cloned lines to a limited extent.

As expected from previous studies, ThId by themselves do not induce an anti-PC antibody response in the presence or absence of specific antigen. However, when combined with a source of ThMHC, they augment the T15+ antibody forming cell response selectively, without affecting the T15- component of the response. This augmenting effect is not MHC restricted to either the B cell or the ThMHC. In limiting dilution assays, ThId selectively increase the number of precursor B cells bearing T15 that can be activated by ThMHC cloned lines, and also lead to an increased burst size of T15+ precursor B cells. The mechanism by which ThId increase the observed precursor frequency in limiting dilution assays is not known. However, as our previous analysis demonstrated that T15+ B cells tend to have a low density of surface Ia glycoproteins, it is tempting to speculate that ThId interact with T15+ B cells by recognition of surface T15 molecules, release factors that induce increased levels of surface Ia molecules, and thus render the B cell activatable by Type 1 ThMHC. This proposed mechanism of action can now be examined in vitro using the cloned ThId lines.

We believe that it is important to distinguish ThId, as described above, from the MHC-restricted, idiotype specific Th that can be raised by immunization with idiotype in adjuvant. Indeed, such cells do not bind idiotype directly, are under conventional Ir gene control,[37] interact with the B cell in a MHC-restricted fashion,[38] are present in normal or increased numbers in mice treated from birth with anti-μ antibody (Rohrer, Kemp and Janeway, unpublished results) or in mice neonatally suppressed for idiotype expression,[39] and require neither antigen nor other types of Th to mediate their effects.[40] The role of this latter type of Th, which has been described in several systems, in antibody responses to antigen is not known at the present time.

Direct Receptor:Receptor Interactions between Cloned ThMHC and Complimentary B Cells Can Lead to B Cell Activation

The exact nature of ThId is still not clear. An analysis of known T cell receptor gene rearrangements may reveal whether such cells express the commonly accepted heterodimeric T cell receptor, or have evolved a distinct set of receptor genes, as has also been proposed for suppressor T cells.[41] In any case, there is evidence that B cells play a role in the definition of the mature T cell repertoire,[42,43] and that such B cell-dependent T cells, especially ThId, in turn influence the development and expression of the mature B cell repertoire (see below).

We have explored the possibility that conventional MHC-restricted helper T cells could interact directly with B cell receptors in ways that might influence the expression of receptors in both sets of cells. To do so, we have taken advantage of an unusual cloned helper T cell line, D10.G4.1, which can be activated by nanogram concentrations or anti-receptor monoclonal antibody in soluble form.[23] We noted that immunization of mice and rats gave rise to a high frequency of anti-T cell receptor monoclonal antibodies, about one hybridoma in twenty having this specificity. This remarkable result suggested to us that D10.G4.1 cells acted as helper cells in their interaction with B cells bearing complimentary receptors. That this was likely to be so was shown both by producing such antibodies in nude mice, and by culturing primed B cells with D10.G4.1 cells. The frequency of anti-receptor monoclonal antibodies in the nude mice was also one in twenty hybrids, and in vitro culture gave rise to anti-receptor nature of such antibodies was confirmed by inhibition of their action with an anti-receptor Fab that inhibits stimulation by anti-receptor antibodies directed at the antigen-recognition region of the receptor. It is of interest that all of the antibodies can be inhibited by this Fab fragment, while the rat monoclonal antibody KJ-16,[44] directed at a framework region of members of a V_B family expressed on D10.G4.1, is not inhibited in its stimulation of D10.G4.1 by this same Fab fragment. Thus not only does D10.G4.1 directly interact with and stimulate B cells bearing complimentary receptors, but these recognize primarily or exclusively antibodies directed at the highly variable recognition site on the receptor molecule.

This system allows us to observe the direct interaction of conventional, antigen-specific, Ia-restricted Th with B cells via receptor-receptor interactions. Our interpretation of this finding is that such interactions are frequent, but that they normally do not lead to the extensive activation observed in the D10.G4.1 system. The reasons for this are as follows. First, D10.G4.1 is unusual in that monoclonal anti-receptor antibodies stimulate rather that inhibit the activation of the cloned T cell line. Second, D10.G4.1 is the most potent helper T cell we have observed in over 100 cloned T cell lines, and secretes very large quantities of a variety of stimulatory lymphokines, including IL-4 and BCGF-II. Third, unlike many of our cloned Th lines, D10.G4.1 has no cytolytic activity; thus upon interaction with a B cell bearing complimentary receptors, no lymphokines are released that might serve to suppressor kill the responding B cell.

One might argue that such interactions account for the observations on ThId described above, and indeed they may. However, there are two differences that need to be examined further. One is the observed requirement for antigen in the functioning of ThId. To date this has not been examined in detail with cloned ThId, so it is possible that antigen stimulates yet a third cell in the earlier ThId assays. While this seems unlikely based on kinetic data,[6] it is not impossible. Further testing or cloned ThId should answer this question. Second, D10.G4.1 delivers a complete activation signal to complimentary B cells. This has not been observed with ThId either in populations or as cloned lines. However, this difference could be due to the unusual physiology of D10.G4.1 rather than to a fundamental difference in lineage. Current studies in our laboratories are focused on answering these questions.

DISCUSSION

The studies outlined above strongly suggest that helper T cell:B cell interactions at low concentrations of antigen, such as are likely to occur in vivo, are effectively cognate. That is, the B cell that is activated must both bind the antigen and bear the correct Ia molecule at a reasonable density. However, we have obtained no evidence that the Ia molecule transmits a signal. More rigorous testing of this conclusion is proceeding. Nevertheless, it is our present view that T cell-dependent B cell activation can be understood as proceeding by the following pathway. B cells bind and concentrate antigen by means of their specific Ig receptors, and this antigen is internalized and processed in acidified endocytic vesicles. Such antigen is presented in association with the B cell's Ia glycoproteins on the B cell surface, and stimulates helper T cells which can recognize the antigen:Ia complex. This recognition event both binds the Th cell to the B cell and cross-links the T cell receptor, leading to Th cell activation and the secretion of B cell activation lymphokines. The nature of the lymphokines secreted appears to be the same regardless of the manner of eliciting their secretion; however, we are currently testing the possibility that interleukin 1 could affect the lymphokiness secreted by a given cloned T cell line. Since IL-1 is expressed on macrophage but not B cell membranes (Horowitz, Kaye, and Janeway, unpublished observations), and since IL-1 appears to be required for T cell growth,[23] such a role for IL-1 could control whether a T cell undergoes a proliferative or an effector pathway of activation.

Certain aspects of this model need further examination. For instance, does binding of antigen to the B cell's receptor alter the susceptibility of the B cell to B cell growth and differentiation factors? Does such an interaction raise the level of Ia antigens on the B cell, making the B cell more readily recognized by the helper T cell? Does the binding of L3T4 to a complimentary structure (probably Ia) in the B cell play a role in B cell activation? These questions can now be asked in a variety of ways using D10.G4.1 cloned Th.

We are also persuaded that whether Th:B cell interactions are cognate or non-cognate, the mechanism by which the Th cell activates the B cell will be the same. Thus, there is no biological difference in the two processes. While this needs further testing, it leads us to ask the following question: Is the cognate behavior of our Th:B interaction assays a result of proximity of the Th cell to the B cell that bears the activating antigen:Ia complex, or does it result from a polar release of B cell activating lymphokines by the Th cell? By this, we mean that the interaction of the Th receptor, and perhaps other structures such as L3T4 as well, with the activating B cell, causes the Th cell to adopt a polarized organization, and to secrete its effector lymphokines exclusively or primarily over that portion of the membrane in direct contact with the B cell bearing the activating antigen:Ia complex. We favor the latter hypothesis for the following reasons. First, it would account for the highly cognate nature of the interactions we observe in an intellectually satisfying manner. Second, morphological correlates for such an hypothesis have been observed with cytolytic T cells,[45] and preliminary observations (by AK) clearly demonstrate that D10.G4.1 cells undergo such cell. Third, we have observed an alteration in the killing of attached targets by drugs that alter microtubule assembly under conditions in which killing of bystander targets was not affected, thus suggesting that such drugs affect the delivery but not the production of effector lymphokines. We are presently testing this hypothesis in both morphological and functional studies.

An alternative or additional hypothesis to account for the cognate behavior of Th:B cell interactions would be that certain B cell activating lymphokines are extremely labile, and can thus travel only very short distances before losing activity. Such putative factors would be difficult or impossible to isolate by procedures normally used to prepare lymphokines. There may be a precedent for such factors in the perforins secreted by cytolytic T cells,[46] which are inactivated rapidly by serum. Finally, the precise sequence in which B cell activating lymphokines are released may be very important in B cell activation. If this is the case, then attempting to mimic direct Th:B interactions with Th supernatants may not be feasible. Such considerations should be borne in mind in attempting to identify such substances.

In addition to conventional Th, our studies clearly identify direct receptor:receptor interactions between Th and B cells. There appear to be three distinct classes of Th:B interactions mediated by receptor:receptor interaction, of which the ThId appear to be the most important, in that one observes ThId function in responses to conventional antigens in all systems which we have examined so far. The function of ThId appears to be to partially activate B cells by means of interactions with the B cell's receptor, rendering such B cells susceptible to conventional Th effects. This is clearly seen in limiting dilution B cell assays. However, the mechanism of this effect is not yet known. We speculate that one mechanism that makes sense is that ThId cause increased Ia antigen density on B cells, as this is a critical parameter of Th:B cell interaction, and as idiotype-bearing B cells in the T15 system at least have been shown to have a low density of surface Ia molecules. Now that cloned putative ThId cells have been prepared, we can ask this question more directly.

The receptor:receptor interactions occuring between Th and B cells could account for the many observations of an influence of B cells on the development of the T cell repertoire. What is lacking in such studies at the present time is a technique for examining the T cell repertoire directly. What is clearly needed are reagents that detect recurrent T cell idiotypes, and a method to analyze their prevalence amongst populations of T cells by direct detection. The indirect functional assays employed in such studies to date, while yielding interesting results, are unlikely to allow us to proceed rapidly to a quantitative estimate of the importance of such influences on the T cell repertoire.

SUMMARY

Helper T cells activate T cells by secreting lymphokines that act on the B cell. Our studies reveal no role for Ia glycoproteins in T cell signaling of the B cell. However, Th:B interaction is functionally cognate, and the questions that now need to be answered are: first, what are the lymphokines that are required to drive a resting, antigen binding B cell to Ig secretion; second, does the binding of antigen by the Ig receptor send an important signal to the B cell or does it serve solely to concentrate the antigen on specific antigen binding B cells; and third, are B cell activating lymphokines released randomly over the surface of the Th cell, or does the Th cell develop a polar orientation upon interaction with the antigen-bearing B cell?

Our studies have also pointed to an important role for idiotype-specific helper T cells in B cell activation. B cells bearing idiotype frequently fail to respond to conventional Th unless they also receive a signal from ThId cells. The precise nature of ThId cells, their relationship to conventional, antigen:Ia specific Th cells, and the mechanism by which they allow idiotype-Bearing B cells to become accessible to conventional Th cells all need to be clarified.

Finally, we describe three types of Th that can potentially interact with B cells via receptor:receptor interactions, one of which is the ThId. The importance of such interactions in the development of the T cell repertoire, suggested by repertoire shifts in B cell deficient mice, and the importance of such B cell dependent T cells on the expression of the B cell repertoire, remains to be determined.

ACKNOWLEDGEMENTS

Many colleagues have provided data, reagents, and ideas to this work. In particular, D. Murphy, P. Flood, A. Bothwell, M. Iverson, and N. Ruddle, our colleagues at Yale, and J. Singer at UCSD have played an important role on both our thinking and our experiments. These studies were supported by NIH grants CA-38350, CA-29606, AI-4579, and AI-13766, and NIH training grant AI-07019, and by the Howard Hughes Medical Institute.

REFERENCES

1. H.N. Claman, E.A. Chapron and R.F. Triplett, Proc. Soc. Exp. Biol. Med. 122:1167 (1966).
2. N.A. Mitchison, Eur. J. Immunol. 1:18 (1971).
3. D.H. Katz, T. Hamaoka and B. Benacerraf, J. Exp. Med. 137:1405 (1973).
4. R.W. Dulton, R. Falkoff, J. Hirst, M. Hoffman, J. Kappler, J.W. Kettman, J.R. Lesley and P. Vann, Prog. Immunol. 1:355, (1971).
5. A. Singer and R.J. Hodes, Ann. Rev. Immunol. 1:211 (1983).
6. C.A. Janeway, Jr., J. Immunol. 114:1394 (1975).
7. C.A. Janeway, Jr., J. Immunol. 114:1408 (1975).
8. C.A. Janeway, Jr., D.L. Bert and F.W. Shen, Eur. J. Immunol. 10:231 (1980).
9. C.A. Janeway, Jr., R.A. Murgita, F. Weinbaum, R. Asofsky and H. Wigzell, Proc. Nat. Acad. Sci. USA 74:4582 (1977).
10. K. Bottomly and D.E. Mosier, J. Exp. Med. 150:1399 (1979).
11. K. Bottomly, C.A. Janeway, Jr., B.J. Mathieson and D. Mosier, Eur. J. Immunol. 10:159 (1980).
12. E.B. Dunn and K. Bottomly, Eur. J. Immunol. 15:687 (1985).
13. K. Bottomly and P.W. Maurer, J. Exp. Med. 152:1571 (1980).
14. K. Bottomly and D.E. Mosier, J. Exp. Med. 154:411 (1981).
15. B. Jones and C.A. Janeway, Jr., Nature 292:547 (1981).
16. K. Bottomly, B. Jones, J. Kaye and F. Jones, III, J. Exp. Med. 158:265 (1983).
17. D.E. Mosier and A. Feeney, J. Exp. Med. 160:329 (1984).
18. L.A. Matis, P.P. Jones, D. Murphy, J. Hedrick, E. Lerner, C.A. Janeway, Jr., J. McNicholas and R. Schwartz, J. Exp. Med. 155:508 (1982).
19. L.A. Matis, L.H. Gilmcher, W.E. Paul and R.H. Schwartz, Proc. Nat. Acad. Sci. USA 80:6019 (1983).
20. P.J. Conrad and C.A. Janeway, Jr., Immunogenetics 20:311 (1984).
21. K. Bottomly and F. Jones, III, In: "B Lymphocyte in the Immune Response," N. Klinman, D.E. Mosier, I. Sher and E.S. Vitetta, eds., Elsevier-North Holland, New York (1981).

22. J.P. Tite, J. Kaye and B. Jones, Eur. J. Immunol. 14:553 (1984).
23. J. Kaye, B. Jones and C.A. Janeway, Jr., Immunol. Rev. 81:39 (1984).
24. R. Zinkernagel and P.C. Doherty, Nature 248:701 (1974).
25. J. Tite and C.A. Janeway, Jr., Eur. J. Immunol. 14:878 (1984).
26. J.P. Tite and C.A. Janeway, Jr., J. Mol. Cell. Immunol. 1:253 (1984).
27. J.P. Tite, M.B. Powell and N.H. Ruddle, J. Immunol. 135:25 (1985).
28. B. Jones, J.P. Tite and C.A. Janeway, Jr., J. Immunol. In press.
29. N.H. Ruddle and B.H. Wakesman, J. Exp. Med. 128:1267 (1968).
30. J. Quintans and R.F. Dick, J. Immunol. 131:609 (1983).
31. K. Bottomly, J. Kaye, B. Jones, F. Jones, III and C.A. Janeway, Jr., J. Mol. Cell Immunol. 1:42 (1983).
32. Y. Asano and R.J. Hodes, J. Exp. Med. 158:1178 (1983).
33. S. Friedman, D. Sillcocks A. Rai, S. Faas and H. Cantor, J. Exp. Med. 161:785 (1985).
34. D.R. Green, P. Flood and R.K. Gershon, Ann. Rev. Immunol. 1:439 (1983)
35. J. Kim, A. Woods, E. Becker-Dunn and K. Bottomly, J. Exp. Med. 162:188 (1985).
36. E.B. Dunn, J. Kim and K. Bottomly, J. Mol. Cell. Immunol. Manuscript submitted for publication.
37. T. Jorgensen and K. Hannestad, J. Exp. Med. 155:1587 (19892).
38. D.J. Jawahana, P. Marrack and J. Kappler, J. Fed. Proc. 41:366 (1982).
39. N. Sakato, C.A. Janeway, Jr. and H.N. Eisin, Cold Spr, Harb. Synp. Quart. Prot. 41:719 (1977).
40. K. Eichmann, I. Falk and K. Rajewsky, Eur. J. Immunol. 8:853 (1978).
41. C.A. Janeway, Jr., J. Mol. Cell. Immunol. 2:57 (1985).
42. C.A. Janeway, Jr., in: "The Biology of Idiotypes," M.O. Green and A. Nisonoff, eds., Plenum Publishing Corp., New York (1984).
43. M-S. Sy, A. Lowry, K.T. Hay Glass, C.A. Janeway, Jr., M. Gurish, M.I. Greene and B. Benacerraf, Proc. Natl. Acad. Sci. 81:3846 (1984).
44. K. Haskins, C. Hannum, J. White, N. Roehm, R. Kubo, J. Kappler and P. Marrack, J. Exp. Med. 160:452 (1984).
45. A. Kupfer, G. Dennert and S.J. Singer, J. Mol. Cell. Immunol. 2:37 (1985).
46. E.R. Podack, Immunology Today 6:21 (1985).

ANTIGEN PROCESSING AND PRESENTATION BY B CELLS

Howard M. Grey, Robert Chesnut, and Jeffrey Krieger

Division of Basic Immunology, Department of Medicine
National Jewish Center for Immunology and Respiratory
Medicine, 1400 Jackson Street, Denver, Colorado

Interest in the question of whether B cells can present antigen to T cells emanates from a consideration of two findings in relation to the immune response to T dependent antigens. The first observation showed that the antigenic determinants recognized by T cells and B cells must be present on the same macromolecule in order for effective antigen-specific T cell help to be delivered to B cells (1). The second important observation to be considered for T-dependent responses was that the accessory cells had to "process" the antigen and display it in the context of the MHC encoded Ia molecules for the recognition by and activation of helper T cells (2,3). Previous to these findings the phenomenon of hapten-carrier linked recognition of antigen by T and B cells had been interpreted as strong evidence in support of the concept that in order to obtain effective T-B cooperation, antigen had to serve as a bridging unit to allow the physical interaction between carrier specific T cells and hapten-specific B cells and that this interaction was required for an effective immune response (1). However,. since in order for T cells to recognize and respond to antigen, the antigen must be processed and presented by an accessory cell in the context of MHC gene products, the postulate that antigen forms a simple bridge between a carrier-specific T cell receptor and a hapten-specific B cell receptor became untenable. This apparent discrepancy between the concept of antigen bridging for effective T-B interactions and the requirement of accessory cell processing and presentation of antigen for T cell recognition would be reconciled if B cells could serve as an antigen presenting accessory cell. The purpose of this paper is to review the evidence in favor of the concept that B cells can in fact serve as antigen presenting accessory cells to T cells of the helper/inducer subset.

Antigen Presentation by B cells

The efforts of our laboratory to test the hypothesis that B cells could present antigen, focused initially on the unique ability of B cells to bind antigen via their surface Ig. We began by utilizing rabbit anti-mouse Ig (RAMIG) antibodies as the antigen (4). RAMIG had two important advantages over conventional antigens for assaying the ability of B cells to present antigen. First, RAMIG binds polyclonally to B cells irrespective of the antigen specificity of their receptor (in contrast to the very low frequency of antigen-specific B cells which, for any given antigen, would make detection of antigen presentation very difficult).

Table 1. Presentation of rabbit anti-mouse immunoglobulin to rabbit IgG-primed T cells by macrophages and B cells

| Antigen Presenting Cells | Antigen | ^{3}H-thymidine incorporation (E-C; cpm x 10^{-3}) in Expt. No. | | | |
		1	2	3	4
Macrophages	NRGG	124	234	121	142
Macrophages	RAMIG	99	186	103	144
B cells	NRGG	0	0	4	0.5
B cells	RAMIG	169	68	148	32

^{3}H-thymidine incorporation expressed as experimental (E) cpm minus unstimulated control (C) cpm.

Second, normal rabbit gamma globulin (NRGG), which would be used to prime the T cells in vivo and which should be presented as effectively as RAMIG by macrophages, would not bind to B cell Ig and thereby could serve as a control antigen to detect macrophage contamination of the B and T cell preparations.

B cells, prepared from the spleens of normal mice, were cultured with T cells isolated from the lymph nodes of mice primed with NRGG in the presence of antigen, either RAMIG or NRGG. The results we observed, shown in Table 1, demonstrated that antigen bound via mIg could be presented to T cells as measured by the induction of a T cell proliferative response. Using spleen adherent cells as a source of macrophages and dendritic cells we found that both NRGG and RAMIG were presented with similar efficiency. In contrast, when B cells were used as the antigen presenting cells, RAMIG was presented as effectively by the B cells as it was by macrophages, while NRGG, which could not bind to the surface Ig on B cells, was not presented. As described above, this latter finding demonstrated that it was indeed the B cells which were the antigen presenting cells in this system and not the result of contaminating macrophages.

Although substantial efforts had been made to ensure that contaminating, non-B antigen presenting cells had been eliminated in these experiments, it was important to reproduce these findings using in vitro cloned sources of B cells and T cells so that any possibility of other cells contributing to the response could be eliminated. To this end, a series of experiments were conducted in which cloned in vitro grown B cell lymphoma lines were tested for their ability to present antigen to cloned antigen-specific, Ia-restricted T cell hybridomas which were also propagated in vitro (5,6). The results of these experiments clearly demonstrated the capacity of Ia$^+$ B cell lymphomas to present antigen to the T cell hybridomas in an antigen specific, MHC restricted fashion, thus proving conclusively the capacity of B cells to act as APC.

The studies described above clearly indicate that B cells can take up antigen either by their Ig receptors (in the case of RAMIG) or by other non-Ig mediated means (in the case of B lymphomas) and that both pathways can lead to the appropriate presentation of antigen to T cells. In the context of the proposed role of B cell antigen presentation under physiological conditions, the capacity of B cells to present antigens which do not bind to their Ig could result in the undesirable secretion of antibodies specific for a large variety of antigens unrelated to the antigen being presented. However, because of the relatively large amounts of antigen required to stimulate T cells when B cells not specific for the antigen or when non-B cells are used as APC, together with the likelihood that antigen-specific B cells would be far more efficient than other B cells in taking up antigen, it was thought that there might be considerable quantitative differences in the B cell's ability to present antigen depending upon the means of antigen uptake. To test this, the RGG-primed T cell system was utilized and a comparison was made between the capacity of the Fab'$_2$ of NRGG and that of the Fab'$_2$ of RAMIG to stimulate RGG-primed T cells when activated B cells were used as a source of antigen presenting cells (7). The results of this study are shown in Table 2. When NRGG Fab'$_2$-pulsed B cells were used to stimulate NRGG-primed T cells, it was necessary to pulse the B cells with 1-6 mg of antigen in order to obtain any significant T cell proliferation. In contrast, when RAMIG Fab'$_2$ was used less than 1 µg was needed to stimulate a similar proliferative response. Since the RAMIG preparation used in this experiment was only 8% specific antibody, these data indicate that RAMIG Fab'$_2$, taken up by the surface Ig receptors, was at least 10^4 fold more efficient an antigen than the NRGG Fab'$_2$ that was taken up by the B cells by a non-Ig mediated means.

Antigen Handling by B cells

Since the macrophage has for several years been considered the prototype of an antigen presenting cell it is useful to compare B cells and macrophages with respect to how antigen is handled by these two types of accessory cells. In order for an accessory cell to present antigen to T cells it must first take up that antigen. There are three mechanisms by which antigen uptake can be accomplished. The first involves fluid phase pinocytosis of the antigen without significant binding to the cell suface prior to endocytosis. The second and third mechanisms both involve the binding of antigen molecules to the cell surface but differ in that one mechanism involves antigen binding via specific receptors

Table 2. Relative efficiency of RAMIG and NRGG to stimulate RGG-primed T cells when activated B cells are used as APC

Ag concentration (µg/ml)	T cell stimulation cpm X 10^{-2}	
	NRGG-Fab'$_2$	RAMIG-Fab'$_2$
1	0	20
10	0	200
100	0	420
1000	2	450
6000	15	440

(e.g. Ig receptors for antigen, Fc receptors, C3 receptors, lectins) located on the accessory cell surface while the other is mediated via non-receptor binding macromolecules on the accessory cell surface. Most protein antigens are taken up by this latter "non-specific" mechanism.

When fluid phase pinocytic capacity was compared between normal B cells, B cell lymphomas and macrophages, striking differences were observed when uptake of horseradish peroxidase was used as an indicator of fluid phase pinocytic activity. Normal macrophages and a macrophage-like cell line (P388D1) had a pinocytic rate of approximately $0.5-1.0$ μl/ 10^7 cells/hour. In contrast to this, B cell tumors, although of approximately the same size as the macrophages, were much less active, pinocytosing fluid at a rate of approximately 0.01 μl/ 10^7 cells/hour. Normal B cells were even less active than B cell tumors with a pinocytic rate of approximately one tenth that of the B cell tumors. Similarly, when the adsorption of iodinated protein to the cell surface of B cells and macrophages was compared, striking differences were observed (5). Using the protein ^{125}I-KLH to measure non-receptor mediated binding to the surface of these cells, under conditions which inhibited endocytosis, macrophages were shown to bind approximately 17 ng of protein/10^7 cells, B cell tumors approximately 2 ng/10^7 cells and normal B cells 0.3 ng/10^7 cells. The extent of binding was dependent upon the quantity of protein incubated with the cells and, as had been previously demonstrated for uptake of antigen by macrophages was not a readily saturable process. The relatively low uptake of antigen by B cells compared to that of macrophages, when the antigen used was taken up by either fluid phase pinocytosis or by non-receptor mediated adsorption to the cell surface, stood in striking contrast to the results obtained when a protein capable of binding specifically to the immunoglobulin receptors on B cells was studied. That is, when radiolabeled rabbit anti-mouse immunoglobulin (RAMIG) was used as the protein, compared to macrophages, B cells took up approximately equivalent amounts of this protein by virtue of the capacity of RAMIG to bind to the approximately 10^5 molecules/cell of surface Ig. As mentioned above this is undoubtedly of crucial importance in allowing antigen-specific B cells to specifically take up sufficient amounts of antigen to allow presentation to T cells especially at low concentrations of antigen.

If, as is elaborated below, antigen presentation by accessory cells requires an active processing event which involves the proteolytic degradation of soluble protein antigens into peptides, then the characterization of the capacity of B cells to endocytose and degrade antigens would also be an important function of these cells. Using the technique of proteolytic stripping of surface bound protein some information on the endocytic capacity of B cells can be obtained. This procedure gives semi-quantitative information with respect to the efficiency of endocytosis of adsorbed protein by measuring that portion of the cell associated protein that is protected from the effects of proteolysis presumably due to endocytosis and sequestration of the protein in an intracellular compartment. Using this method, after a 2 hour incubation at 37°, both macrophages and B cells could be shown to internalize approximately 30% of rabbit anti-mouse immunoglobulin bound to their surface. However, antigen (KLH) bound nonspecifically to the surface of B cells was not detectably compartmentalized into a protease resistant site. In contrast, when KLH was bound to the surface of peritoneal macrophages approximately 20% of the protein was internalized during the same time interval. Thus, the extent of sequestration of protein into protease resistant sites depends upon the mode of binding to the B cell surface: rabbit anti-mouse Ig interaction with cell surface Ig leads to efficient internalization whereas antigen binding to nonspecific membrane sites does not.

98

Next, the relative capacity of B cells and macrophages to degrade proteins initially bound to their cell surface was studied by measuring the release of acid soluble radioactivity during a period of incubation at 37°C (Table 3) (8). It is clear from the data in this table that antigens bound to the surface of B cells via non-Ig mediated mechanisms were very inefficiently degraded compared with the fate of the same antigen taken up by macrophages. The rate of protein degradation as measured by the generation of TCA soluble radioactivity was 1-5% of the rate observed in macrophages. In striking contrast to the data obtained with proteins taken up non-specifically by B cells, when the degradation of RAMIG was studied, B cells and macrophages displayed comparable rates of degradation. Thus, it appears that ligand bound to membrane immunoglobulin is preferentially delivered to a site of catabolism where it is degraded at a rate comparable to that of macrophages. Antibodies to other surface antigens such as β_2 microglobulin were degraded at much slower rates, similar to proteins taken up nonspecifically, as were immune complexes bound to B cell Fc receptors. This striking difference in catabolism between Ig mediated and with other mechanisms of uptake of antigen by B cells may be another important factor in determining their preferential capacity to function as efficient APC for T cells that are specific for the same antigen as the B cells.

Earlier studies utilizing macrophages as accessory cells strongly suggested that active metabolic events were required following the binding of antigen to the accessory cell surface, before antigen could be effectively presented to and recognized by T cells. The evidence used to support the active "processing" of antigen included: 1) a finite time of incubation at 37°C after antigen was bound to the cell surface had to elapse before macrophages could present antigen effectively to T cells; 2) inhibitors of energy requiring metabolic events prevented antigen presentation to T cells; 3) antigen became resistant to antibody

Table 3. Degradation of proteins taken up by normal B cells, B cell lymphomas (BAL), and macrophages (PEC)

Experiment	Cell	^{125}I Antigen	^{125}I Protein Bound (cpm x 10^{-3})	% Degraded /hr
1	PEC	KLH	449	40.1
	BAL	KLH	13	<0.1
2	PEC	D-OVA	112	18.6
	BAL	D-OVA	33	1.1
3	PEC	RAMIG	294	13.4
	BAL	RAMIG	136	7.0
	nlB	RAMIG	137	6.5
4	PEC	KLH-anti-KLH	636	8.9
	BAL	KLH-anti-KLH	268	0.1
5	PEC	RAMIG	283	14.4
	BAL	RAMIG	219	8.7
	PEC	anti-β_2m	39	11.9
	BAL	anti-β_2m	39	0.1

blockade or protease stripping of the cell surface after incubation of antigen pulsed macrophages at 37°C for several hours; 4) lysosomotropic amines inhibited antigen presentation by macrophages; 5) inactivation of macrophages by ultraviolet light or aldehyde fixation before exposure to antigen prevented effective antigen presentation even in the presence of added interleukin-1 (IL-1). Such studies suggested that internalization and limited proteolysis of antigen may be necessary events that must occur prior to presentation of antigen by macrophages. Since, as detailed above there are striking differences in the efficiency with which B cells and macrophages bind, endocytose, and degrade antigen, the possibility existed that the requirements for processing of antigen might differ depending upon the accessory cell involved. To examine this, macrophages and B cell lymphomas were tested for their capacity to present antigen (taken up by non-specific means) to antigen-specific T cell hybrids as measured by the production of IL-2 by the hybrids. To study the kinetics of antigen presentation, antigen was allowed to bind in the cold to the surface of the accessory cells followed by washing and incubation at 37°C for varying periods of time, after which the cells were mildly fixed with paraformaldehyde. Both B cell lymphomas and macrophages exhibited similar kinetics in that no antigen presenting capacity was demonstrable prior to 45 minutes at 37° C at which time some minimal activity was observed with B cell tumors which increased over a period of about another hour. Similarly, peritoneal exudate macrophages started to show antigen presenting capabilities at 60 minutes which progressively increased for the next hour.

Somewhat stronger evidence for proteolytic degradation as an important processing event has come from the study of the effect of a variety of weak bases, including chloroquine, on the antigen presenting capacity of accessory cells. These "lysosomotropic" agents have been shown to be concentrated within lysosomes, raise the lysosomal pH and, as a consequence, to inhibit the function of a variety of acid hydrolases contained within this organelle. When chloroquine was added prior to and during the time required for antigen processing it was found that both B cell lymphomas and peritoneal exudate macrophages were similarly inhibited in their capacity to present antigen to T cell hybrids in a dose dependent fashion with concentrations $\geqslant 40$ µ M being completely inhibitory (9). In other experiments this inhibition was demonstrated to be restricted to the processing events rather than having an affect on antigen display, since cells that were pulsed with antigen prior to chloroquine treatment showed a normal capacity to present that antigen.

With several of the T cell hybridomas that were used for the studies described above, it has been possible to define the processing event more definitively as involving proteolytic degradation. The results of this work demonstrated that accessory cells, either macrophages or B cells, that were fixed with paraformaldehyde or glutaraldehyde prior to exposure to antigen, while they were totally incapable of presenting intact protein antigen to the T cell hybridomas, were very efficient in presenting certain peptides derived from that antigen that had been generated by either enzymatic or chemical cleavage of the antigen prior to its being added to the fixed presenting cells (10,11). One example of this type of experiment is shown in Table 4 which demonstrates that both peritoneal macrophages and B cell lymphomas which had been prefixed before addition to the culture had a similar capacity to present a peptide of ovalbumin to the 3D054.8 hybridoma; whereas both cell types when prefixed failed to present the intact native or denatured antigen. Although the studies shown in this table were performed with a B cell lymphoma, identical results have been obtained with normal mitogen-activated B cells.

Table 4. Capacity of pre-fixed B cells and macrophages to present OVA peptides to OVA specific T cell hybrids

Accessory cells	Antigen (units/ml IL-2 production)			
	N-OVA	D-OVA	OVA Tryptic peptides	OVA CNBr peptides
Live macrophages	160	160	320	320
Fixed macrophages	<20	<20	640	320
Live B lymphoma (A20)	1280	1280	640	1280
Fixed B lymphoma (A20)	<20	<20	320	320

In conclusion, when macrophages and B cells are compared by functional criteria, they behave in precisely the same manner to all the manipulations that have been used to support the concept of active metabolic processing events being a necessary aspect of antigen presentation to T cells. In rather striking contrast to this is the comparison between B cells and macrophages when the handling of antigen is examined biochemically. Whereas macrophages bind, endocytose and degrade protein antigens very efficiently, B cells perform the same functions 10-100 fold less efficiently. The one important exception to this statement is the efficient manner in which RAMIG is taken up and degraded by B cells. As mentioned above this exception may be of considerable importance in permitting antigen-specific B cells to efficiently process and present the antigen for which its mIg has specificity.

Antigen Presenting Capacity of Resting and Activated B Cells

The question of whether resting B cells can effectively present antigen to helper T cells or whether they must be activated before being capable of carrying out this function remains an important issue with regard to T-B collaboration and the delivery of helper signals to the B cell. Concern over this originally surfaced as a result of studies that investigated the ability of normal B cells and B cell tumors to present antigen to KLH-specific T cells (5). Somewhat surprisingly it was found that while B cell lymphomas and mitogen activated B cells were quite effective at presenting KLH to the T cell hybridoma, normal B cells showed no antigen presenting capacity except at the very highest cell numbers tested and appeared to be at least 50 fold less efficient than the activated B cells or B lymphoma cells.

Subsequent to these findings, Ashwell and co-workers published results showing that the antigen presenting function of normal B cells was diminished by exposure of the B cells to gamma irradiation and that if high dose (>1000 rads) irradiation was avoided small resting B cells appeared to be as efficient as macrophages/dendritic cells at presenting antigens to T cells (12). Since we and other workers had used relatively large doses of gamma irradiation (3,000-4,000 rads) to block B cell proliferation prior to their being used as antigen presenting cells, Ashwell et al. reasoned that the previous inability to observe efficient antigen presentation by small resting B cells was probably a result of the affect of the irradiation on B cell function.

In order to investigate the relative capacity of resting vs activated splenic B cells more extensively, our laboratory initiated a series of studies in which lightly irradiated (500 rads) splenic B cells were separated into four fractions on the basis of their buoyant density using Percoll gradient centrifugation (13). Cells isolated from 50/60%, 60/65%, 65/72% Percoll interfaces and a >72% pellet, were tested for their accessory cell function in 3 different assay systems, each making somewhat distinct demands on the accessory cells employed. The assay systems used included: 1) The response of the I-A^d-restricted ovalbumin (OVA)-specific T cell hybridoma, 3D0-54.8 to intact native OVA; presentation which requires antigen uptake and processing prior to display of the appropriate epitope to the T cell. 2) The response of 3D0-54.8 to a tryptic-digest of OVA which had been previously shown to contain the OVA peptide 323-339 which needs only to bind appropriately to the B accessory cell but does not require antigen processing prior to presentation (10,11). 3) Con A stimulation of unprimed splenic T cells; a system in which there is neither genetic restriction nor a requirement for processing but which may require the elaboration of a co-stimulator activity such as IL-1.

Typical results are shown in Figure 1 in which varying numbers of B cells from each of the fractions were tested for their accessory cell function. Although somewhat different quantitatively, the results obtained with the 3 assay systems were qualitatively quite similar in that B cells from the lowest density fraction provided the greatest accessory cell function, and that function declined progressively with increasing B cell density such that the B cells which passed through the 72% Percoll layer expressed little or no accessory cell function at any of the B cell concentrations tested. Thus these data suggest that even at low doses of irradiation resting B cells are quite inefficient accessory cells.

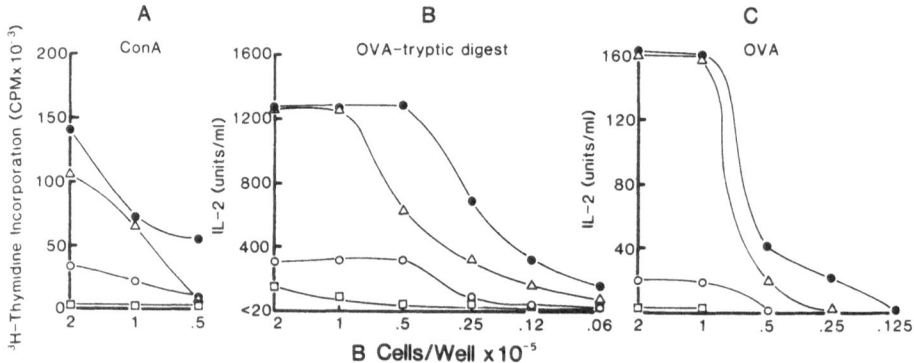

Figure 1.   Accessory cell function of Percoll fractionated splenic B cells. B cells obtained from the 50/60% (●), 60/65% (Δ), 65/72% (o), and >72% (□) fractions were used as accessory cells for a Con A-mediated splenic T cell proliferative response (panel A); as antigen-presenting cells for the presentation of OVA-tryptic digest (panel B); or antigen-presenting cells for the presentation of OVA (panel C). For panels B and C, the I-A^d-restricted, OVA-specific T cell hybridoma, 3D0-54.8 was used. B cells were exposed to 500 rad of γ-irradiation before use.

The reason why activated B cells are much more effective accessory cells than resting B cells remains controversial. Several possibilities exist for which data has been published; each of which, depending upon the nature of the antigen and the T cell population tested may contribute at least in part to the enhanced capacity of activated B cells to present antigen. One property which might affect the relative capacity of resting vs activated B cells to present antigen could be their ability to bind antigen. As described above, it had been previously shown that normal B cells and B cell lymphomas differed in their capacity to bind and degrade intact protein antigens and that this difference appeared to correlate with their relative antigen presenting capacity. However, it was not possible to determine from these studies if differences in antigen presenting capacity were the result of differential antigen binding or differences between the various populations of B cells in their ability to process (i.e., internalize and degrade) and reexpress the processed antigenic moiety on the B cell surface. To evaluate this question more directly the differences between normal B cells, activated B cells and B lymphoma cells in their ability to bind and present an antigenic 17 amino acid OVA-peptide was examined. The results of these experiments showed that at any given peptide concentration, B lymphoblasts and B lymphoma cells were able to bind approximately 5 fold more peptide than resting B cells. When a similarly pulsed population of resting B cells, activated B cells and B lymphoma cells were tested for their ability to present the OVA-peptide to an OVA-specific T cell hybridoma, it was found that the activated B cells and A20-1.11 cells were efficient APC and stimulated 640 units/ml IL-2 while normal B cells stimulated only 40 units/ml IL-2. Thus, at least for this peptide, there was a direct correlation between the capacity of resting and activated B cells to bind the peptide and their capacity to functionally present the peptide to a T cell hybrid.

Another possible explanation for the differential capacity of resting B cells, activated B cells and macrophages to present antigen has to do with the structure of the Ia molecule itself. While high pressure liquid chromatography analysis has shown that the Ia molecules synthesized by spleen adherent cells and B cells have apparently identical polypeptide structures isolectric focusing studies have indicated the presence of differences which appear to be due to varying degrees of sialylation of the Ia α chain; the α chain from resting B cells having greater quantities than that from macrophages or activated B cells (14). Frohman and Cowing (15) have recently published results showing that Percoll fractionated splenic B cells enriched for resting B cells are approximately 250 times less efficient than LPS-activated B cells at presenting KLH to KLH primed T cells. However, when the resting B cell fraction was treated with C. perfringens neuraminidase to remove sialic acid, a 24 fold increase in the antigen presenting function of resting B cells was observed. More information on this interesting neuraminidase effect is required, however, before it can be concluded that the effect is mediated by removal of sialic acid from Ia molecules on resting B cells.

Another change in Ia that occurs following B cell activation that might play a role in the enhanced APC function of activated B cells relates to the quantity of Ia expressed on resting and activated B cells. Several groups have reported that activation of B cells results in a 2-10 fold increase in Ia expression. How the level of Ia expressed on accessory cells might influence APC function has been examined by several groups of investigators. Matis and co-workers demonstrated that under conditions of limited antigen concentration, the number of Ia molecules expressed on the antigen presenting cell directly influenced the magnitude of the T cell response (16). The presentation of low concentrations of pigeon cytochrome C to an I-E^k-restricted cytochrome C-specific T cell clone by splenic antigen presenting cells expressing 1, 1/2 or 1/8 the level of I-E^k resulted in proliferative responses that were inversely proportional to the level of Ia expression. It was further found that the low expression of Ia could be compensated for by a proportional increase in the antigen concentration.

The problem in interpreting such experiments is that while it is evident that there are differences in the level of expression of Ia on the various antigen presenting B cell populations, it is impossible to determine if there are concomitant differences, either qualitative or quantitative in the expression of non-Ia molecules which may also be important for T cell interaction and for which we do not yet have a means of detection. In an effort to circumvent these problems our laboratory has utilized the recently developed technique of constructing an artificial antigen presenting surface composed of purified Ia molecules incorporated into a cholesterol and phospholipid matrix. Liposomes prepared with these lipids and Ia can be coated onto glass beads and used to present the previously described ovalbumin peptides to Ia-restricted ovalbumin-specific T cell hybridomas. By varying the amount of Ia incorporated into the synthetic membrane and the amount of antigenic peptide added to the culture the influence of Ia density and antigen concentration can be evaluated independent of other macromolecules. Shown in Table 5 is an example of the results that were obtained in carrying out such a study. The amount of peptide antigen required to induce a detectable quantity (20 units) of IL-2 by the responding T cell hybridoma was determined for beads coated with varying amounts of Ia. The results indicate that the amount of antigen required to elicit an IL-2 response is inversely proportional to the Ia density so that the product of the Ia concentration and antigen concentration remained relatively constant over the range of Ia concentrations tested (last column). This finding is a direct confirmation of the results of the studies described above using intact cells as the antigen presenting surface and demonstrates that when the concentration of antigen is limited, the density of Ia molecules alone can significantly affect the ability of a cell to present antigen. Our laboratory has recently extended this type of experiment to explore the defect in the antigen presenting cell function of resting B cells. Shown in Figure 2 are the results obtained when the I-A^d molecules isolated from normal splenic B cells were compared with the I-A^d isolated from the B cell lymphoma, A20-1.11, and the I-A^d present on intact fixed A20-1.11 cells. The results are plotted as the product of I-A expressed on the liposome coated glass beads or A20-1.11 cells (i.e., molar concentration of IA) and the molar concentration of antigenic peptides vs the quantity of IL-2 (log$_2$) produced by the responding T cell hybridoma. By plotting the product of the Ia and antigen concentrations the inverse relationship between antigen and Ia concentration described above were normalized thus allowing a direct comparison of the efficiency of the I-A^d molecules from different sources in mediating antigen presentation. The data illustrate two points. First, there is approximately a 10-20 fold difference in the ability of Ia isolated from normal splenic B cells, and the Ia isolated from A20-1.11 cells, in eliciting an IL-2 response from the OVA specific T cell hybrid. Although there are certainly other possible explanations for these results they suggest that there is a structural difference between the Ia from normal B cells and the B lymphoma cells which allows the latter to be more effective at presenting antigen. Such structural differences could be related to the carbohydrate structure on the Ia molecules as suggested by the studies of Cowing and coworkers, or other differences in the protein structure which have thus far gone undetected using monoclonal antibodies and two dimensional gel electrophoresis analysis.

Table 5. Relation between Ia and peptide concentration
required to stimulate T cells

Ia (ng/ml)	peptide (ng/ml)	[Ia] X [peptide]
5.6	100	560
9.6	75	720
22.4	25	560
36.8	13	480

The second piece of information illustrated by this figure has to do
with the relative effectiveness of Ia bearing liposome coated glass beads
and an intact fixed cell at presenting antigen. When fixed A20-1.11
cells were compared to the I-A which had been isolated from A20-1.11
cells and incorporated into liposomes, the I-A expressed on intact A20
cells was approximately 20-30 fold more effective at eliciting an IL-2
response than the purified A20-1.11 I-A expressed on liposome coated
glass beads. These results suggest that there are characteristics about
the antigen presenting cell surface other than Ia structure which dramatically enhance the ability of A20-1.11 cells to act as an antigen
presenting cell. These differences could be related to differences in
the mobility of Ia molecules in the plane of the membrane on intact cells
which may be very different than that of Ia molecules that are expressed
in the synthetic lipid bilayer; alternatively other molecules expressed

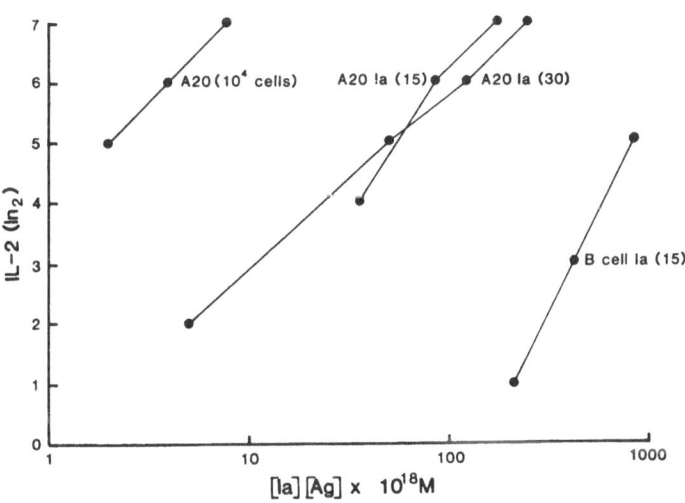

Figure 2. A comparison of the antigen presenting capacity of liposomes containing I-A isolated from normal B cells or A20
B lymphoma cells vs intact fixed A20 B lymphoma cells.
The numbers in parentheses represent two different bead
preparations constructed with liposomes formed with two
different concentrations (15-30 μg/ml) of Ia.

on accessory cells but not present in the liposomes which have yet to be identified, may also be involved in enhancing the effectiveness of intact cells. If these auxilary molecules exist it is possible that activation of B cells could enhance their expression which could also increase the antigen presenting function of activated B cells compared to resting B cells.

Another possible factor responsible for the difference in accessory cell function between resting and activated B cells could be the differential capacity to provide a costimulator, IL-1-like signal to the responding T cells. While the exact role(s) of IL-1 in T cell activation remains controversial it does appear to function in the elaboration of IL-2 and, in addition, may influence the expression of IL-2 receptors on the T cell surface. Although efforts by our laboratory as well as those of other investigators to detect the presence of IL-1 activity in the culture supernatant of either resting or activated murine B cells or B cell lymphomas have failed to detect such an activity, Oppenheim and his coworkers have recently reported the presence of an IL-1 like activity produced constituitively by EBV transformed human B lymphocytes and by normal human peripheral blood B lymphocytes following stimulation with lipopolysaccharide (17). Recent experiments by Kurt-Jones et al. (18) have suggested that murine B cells may express a membrane bound form of IL-1-like activity similar to that previously described by these workers in macrophages. Using an IL-1 dependent T cell line (D10.G4.1) these workers found that B cells which had been activated with mitogens plus T lymphokines could support the Con A-stimulated growth of D10.G4.1 cells presumably by providing a source of membrane bound IL-1 like activity.

Studies in our laboratory concerning the expression of an IL-1-like signal by B cells has centered on the capacity of B cells to stimulate an allogeneic primary mixed lymphocyte response (MLR). While paraformaldehyde fixed-normal B cells, were able to stimulate a secondary MLR or allo-Ia-specific T cell hybridomas they were unable to stimulate a primary MLR. However, the capacity of B cells to stimulate a primary MLR could be reconstituted by the addition of macrophage derived IL-1 to the culture. In contrast to normal B cells, B cells stimulated for 3 days with LPS + DexSO$_4$ were able to stimulate a primary MLR suggesting that a co-stimulator activity similar in its action to IL-1 was produced by the LPS and DexSO$_4$ stimulated B cells. However, in contrast to the studies of Kurt-Jones et al., these LPS + DexSO$_4$ stimulated B cells could not support the growth of the IL-1 dependent cell line, D10.G4.1, suggesting that this activity was not due to IL-1 as defined in this system. Clearly much more work is required before any conclusions can be made regarding the relationship between this co-stimulator activity associated with activated B cells and IL-1.

In summary it is clear that B cells share with other MHC class II-expressing cells the capacity to present antigen to T cells. The unique characteristic of B cell antigen presenting capacity lies in the ability of specific B cells to efficiently take up antigen via their immunoglobulin receptors and to present that antigen to antigen-specific helper T cells. It has recently been established that the presentation of antigen by antigen-specific B cells is highly efficient compared to non-Ig mediated antigen presentation by B cells or non-B cells due to the fact that these latter cells do not have the requisite antigen specific receptors to facilitate antigen uptake. Although this proposed mechanism of antigen presentation to T cells by B cells explains many features of T-B interactions, such as the MHC-restriction and the requirement for hapten and carrier to be on the same macromolecule in order to facilitate efficient T-B interaction, the relevancy of this mechanism for an in vivo immune response remains to be demonstrated.

REFERENCES

1. Mitchison, N.A. 1971. The carrier effect in the secondary response to hapten protein conjugates. II cell cooperation. Eur. J. Immunol. 1:68.

2. Rosenthal, A. and Shevach, E. 1973. Function of macrophages in antigen recognition by guinea pig T lymphoctes. J. Exp. Med. 138:1194.

3. Schwartz, R., Yano, A. and Paul, W. 1978. Interactions between antigen-presenting cells and primed T lymphocytes. An assessment of Ir gene expression in the antigen-presenting cell. Immuno. Rev. 40:153.

4. Chesnut, R. and Grey, H.M. 1981. Studies on the capacity of B cells to serve as antigen-presenting cells. J. Immunol. 126:1075.

5. Chesnut, R., Colon, S. and Grey, H. 1982. Antigen presentation by normal B cells, B cell tumors, and macrophages: Functional and biochemical comparison. J. Immunol. 128:1764.

6. Walker, E., Warner, N., Chesnut, R., Kappler, J. and Marrack, P. 1982. Antigen specific, I region-restricted interactions in vitro between tumor cell lines and T cell hybridomas. J. Immunol. 128:2164.

7. Kakiuchi, T., Chesnut, R.W. and Grey, H.M. 1983. B cells as antigen-presenting cells: the requirement for B cell activation. J. Immunol. 131:109.

8. Grey, H.M., Colon, S.M. and Chesnut, R.W. 1982. Requirements for the processing of antigen by antigen-presenting B cells. II. Biochemical comparison of the fate of antigen in B cell tumors and macrophages. J. Immunol. 129:2389.

9. Chesnut, R., Colon, S. and Grey, H.M. 1982. Requirements for the processing of antigens by antigen-presenting B cells. I. Functional comparison of B cell tumors and macrophages. J. Immunol. 129:2382.

10. Shimonkevitz, R., Kappler, J., Marrack, P. and Grey, H.M. 1983. Antigen recognition by H-2-restricted T cells. I. Cell-free antigen processing. J. Exp. Med. 158:303.

11. Shimonkevitz, R., Colon, S., Kappler, J.W., Marrack, P. and Grey, H.M. 1984. Antigen recognition by H-2-restricted T cells. II. A tryptic ovalbumin peptide that substitutes for processed antigen. J. Immunol. 133:2067.

12. Ashwell, J.D., DeFranco, A.L., Paul, W.E. and Schwartz, R.H. 1984. Antigen presentation by resting B cells. Radiosensitivity of the antigen-presentation function and two distinct pathways of T cell activation. J. Exp. Med. 159:881.

13. Krieger, J., Grammer, S., Grey, H. and Chesnut, R. 1985. Antigen presentation by splenic B cells: Resting B cells are ineffective, whereas activated B cells are effective accessory cells for T cell responses. J. Immunol., in press.

14. Cullen, S., Kindle, C., Shreffler, D. and Cowing, C. 1981. Differential glycosylation of murine B cell and spleen adherent cell Ia antigens. J. Immunol. 127:1478.

15. Frohman, M. and Cowing, C. 1985 Presentation of antigen by B cells: Functional dependence on radiation dose, interleukins, cellular activation, and differential glycosylation. J. Immunol. 134:2269.

16. Matis, L., Glimcher, L., Paul, W. and Schwartz, R. 1983. Magnitude of response of histocompatibility-restricted T cell cones is a function of the product of the concentration of antigen and Ia molecules. Proc. Natl. Acad. Sci. USA 80:6019.

17. Matsushima, K., Procopio, A., Abe, H., Ortaldo, J. and Oppenheim, J. 1985. Production of interleukin-1 activity by normal human peripheral blood B lymphocytes. J. Immunol. 135:1132.

18. Kurt-Jones, Kiely, J., and Unanue, E. 1985. Conditions required for expression of membrane IL-1 on B cells. J. Immunol. 135:1548.

THE ROLE OF RECEPTOR IMMUNOGLOGULIN IN ANTIGEN-SPECIFIC, MHC-RESTRICTED T CELL/B CELL COOPERATION

David C. Parker and Hans-Peter Tony*

Department of Molecular Genetics and Microbiology
University of Massachusetts Medical School
Worcester, MA 01605

INTRODUCTION

Clonal selection theory assigns a central role to membrane Ig in determining the specificity of the antibody response. However, the combination of antigen with membrane Ig is insufficient to induce resting B cell to produce a clone of antibody secreting cells; for the

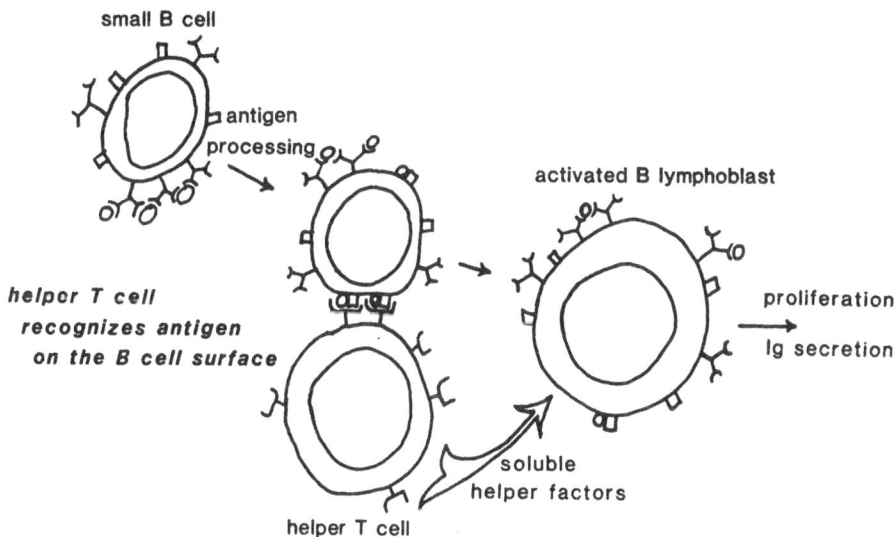

Figure 1. Model of T cell/B cell cooperation.

*Present address: Dr. Hans-Peter Tony, Medizinische Poliklinik der Universität Würzburg, Klinikstrasse 8, 8700 Würzburg, WEST GERMANY

antibody response to soluble protein antigens, the same molecule of antigen needs to be recognized by an antigen-specific helper T cell. We share Howard Grey's perspective presented earlier that B cells get help by acting as very efficient antigen-presenting cells (APC) for antigens that bind to membrane Ig, and that recognition of the same molecule by B cells and T cells is sequential rather than simultaneous. This scheme (Figure 1) explains MHC-restriction of T cell/B cell cooperation, as well as the ability of T cells, which see processed antigen and tend to cross-react extensively on denatured proteins or peptides, to help B cells make antibodies which can be exquisitely sensitive to protein conformation.

Since antigen-specific, T-B cooperation is normally an exceedingly rare event, we have developed a murine, in vitro polyclonal model for a T lymphocyte-dependent antibody response in which rabbit anti-mouse immunoglobulin is used in place of antigen. T cell help is provided by cultured T cell lines and cloned T cell hybridomas which recognize the rabbit anti-immunoglobulin reagents as foreign protein antigens (1). Using this system, we have shown that recognition of anti-IgM or anti-IgD on the B cell surface is restricted by class II products of the major histocompatibility complex (MHC), and results in both a T cell response and a vigorous polyclonal B cell response. This versatile model incorporates the most important features of antigen-specific, MHC-restricted help in a polyclonal response which gives us direct access to otherwise exceedingly rare responding cells and cell interactions. The model extends earlier studies by Chesnut and Grey, who showed that B cells can present rabbit anti-Ig very efficiently to rabbit globulin-primed lymph node T cells (2). We are using the model to sort out the roles of membrane Ig and helper T cells in the antibody response.

ROLES FOR MEMBRANE Ig

There is abundant evidence that cross-linking membrane Ig delivers an activating signal to the B cell (3) which, in the presence of the appropriate T cell-derived helper factors, can result in clonal expansion and Ig synthesis (4). Involvement of membrane Ig in antigen presentation creates new potential roles for membrane Ig in antigen processing and in signalling the B cell to induce a helper signal from the T cell, for example, by induction of increased class II molecule expression (5) or induction of membrane IL-1 (6).

One indication that direct signalling via membrane Ig may not be necessary comes from the very low doses of anti-Ig which are required when help is provided by specific T cells rather than helper factors. Figure 2A shows that a good polyclonal secretory response in the presence of T cells can be obtained with about 1ng/ml of anti-Ig, 10,000-fold less than that required to induce a response with helper factors, and well below the concentration which, by itself, produces any measurable effect on B cells. Figure 2B shows that a rabbit antibody which does not bind specifically to membrane Ig, rabbit anti-phenylarsonate, can induce a response in small B cells. However, since it does not bind specifically to the B cell, it is not presented efficiently, and must be added at 10,000-fold higher concentration.

Giving a putative membrane Ig-mediated signal with very low concentrations of goat anti-Ig antibodies does not alter the response to rabbit anti-phenylarsonate (anti-Ars). However, if the goat anti-Ig antibody is arsanilated, then rabbit anti-Ars can bind to it, and thus, indirectly, to membrane Ig. The dose response then shifts to very low concentrations of rabbit anti-Ars, comparable to those of

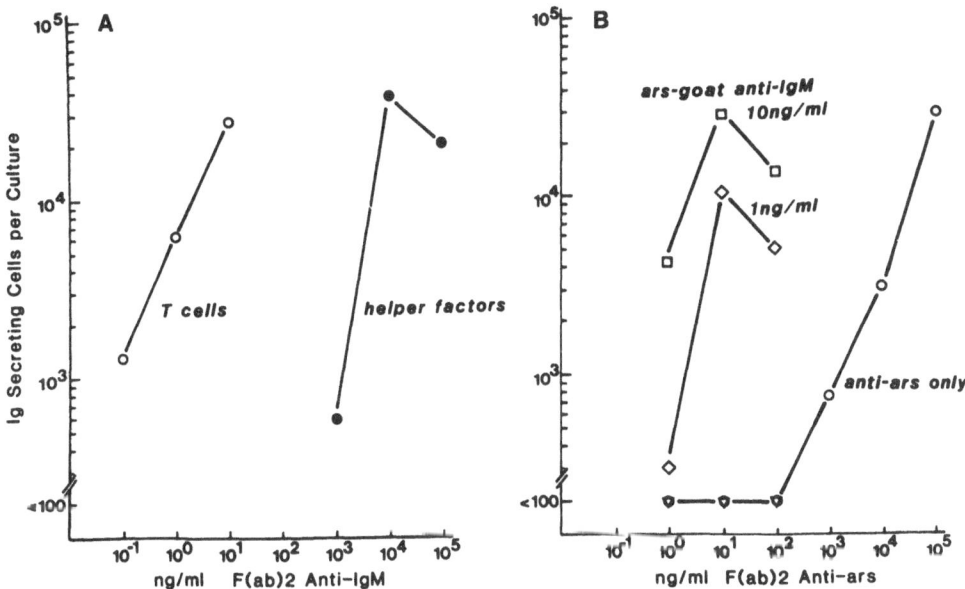

Figure 2. (A) A globulin-specific T cell line (3 x 10^4 CDC35 cells) induces a secretory response from 10^5 small B cells at 10,000-fold lower concentrations of anti-Ig than that required with helper factors. (B) Rabbit $F(ab')_2$ anti-Ars, which does not bind specifically to the B cell, will also induce a response with T cells present, but at much higher concentrations (O). However, ng/ml quantities of $F(ab')_2$ anti-Ars are adequate if linked to the B cell surface by a hapten sandwich with arsanilated goat anti-IgM at 1 ng/ml (◇) or 10 ng/ml (□). 10 ng/ml goat anti-IgM which was not arsanilated (▽) failed to enhance the response to $F(ab')_2$ anti-Ars. From reference #1.

rabbit anti-Ig (Figure 2). This experiment demonstrates an important role for membrane Ig in antigen presentation by B cells, but suggests that a signalling role is not essential if the helper T cell can recognize antigen on the B cell surface. Similar conclusions have been drawn by other laboratories based on responses of B cells to alloreactive T cells (7) or antigen-specific T cells at high antigen concentrations (8,9). Other groups have shown that antigen-specific B cells (10,11) or B cell lines (12) act as incredibly efficient antigen-presenting cells for antigens which they can bind to their endogenous membrane Ig.

Even at these very low concentrations of anti-Ig, efficient presentation of anti-Ig via membrane Ig could well involve a transmembrane signal which could aid in processing and presentation of antigen or in induction of the helper T cell. Since all the direct effects of anti-Ig on B cells depend upon cross-linking the membrane Ig, we compared the efficiency of presentation of divalent Fab_2 anti-Ig with monovalent Fab anti-Ig (13). Figure 3A shows that membrane Ig cross-linking has no effect on antigen presentation, since Fab_2 and Fab anti-Ig are presented with the same efficiency, as

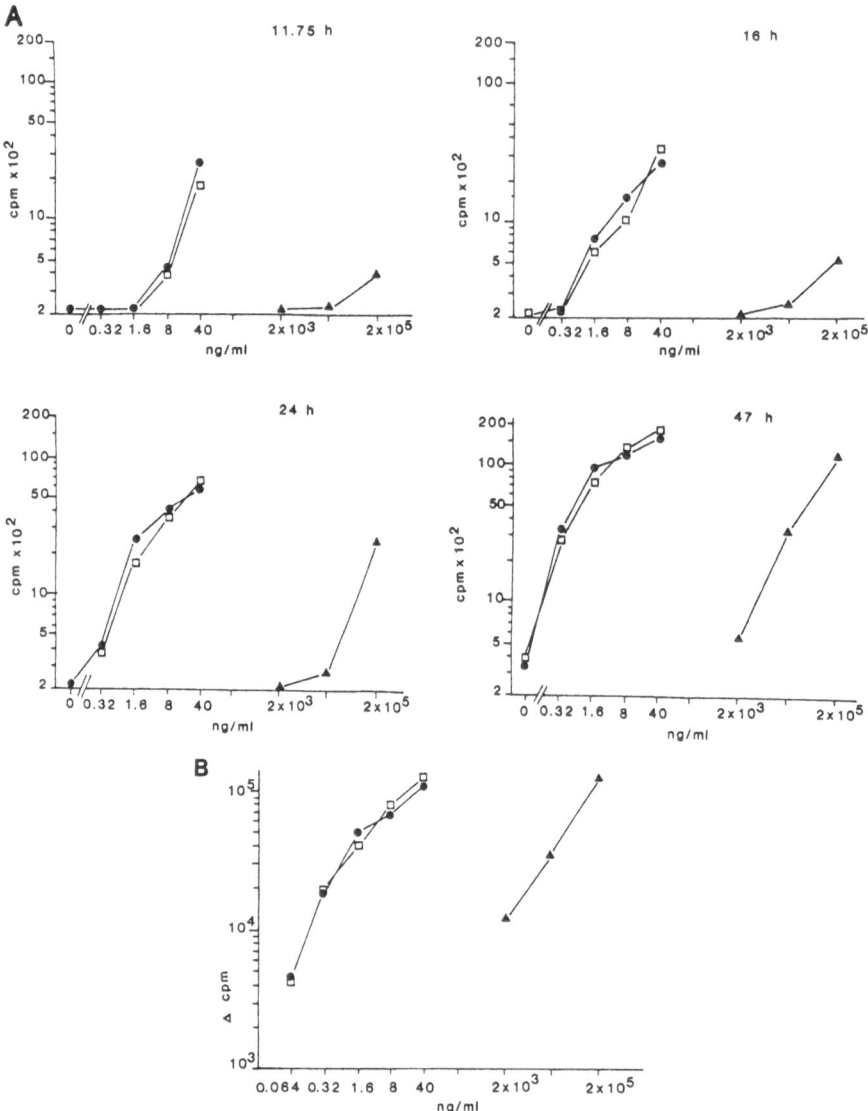

Figure 3. Crosslinking of mIg is not required for either antigen presentation by small B cells for for B cell activation. 10^5 small C3D2 B cells were cultured with 3×10^4 T cells (CDC35). Cultures received the indicated amount of F(ab')$_2$ anti-IgM (●), Fab' anti-IgM (□), or absorbed F(ab')$_2$ NRG (▲). (A) Il-2 accumulation was measured at 11.75h, 16h, 24h, and 47h. (B) [^{3}H]Thymidine incorporation by B cells was measured at 68h. From reference #13.

measured by IL-2 production by the T cells, even at the earliest time point. Moreover, cross-linking appears to be without effect on the B cell response, since the dose response curves for the B cell proliferative response to Fab$_2$ and Fab anti-Ig in the presence of T cells are identical (Fig. 3B). This is not to deny the wealth of evidence that membrane Ig-mediated signals can enhance responses in other experimental systems involving T cell-driven responses (14, 15). It may be that our T cell lines are particularly adept at interacting with small, resting B cells (see below).

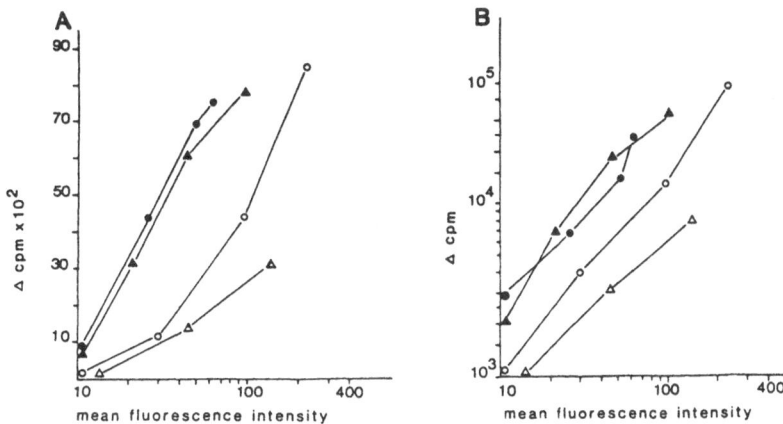

Figure 4. Comparison of antibodies directed to mIg vs. class I MHC molecules for their efficiency in antigen presentation and B cell activation. Small B cells were pulsed at 4° in the presence of sodium azide, and a portion was stained as described below. Another portion of the cells was washed and placed into culture. Cultures contained 10^5 pulsed small C3D2 B cells and 3 x 10^4 irradiated T cells of the line CDC25. IL-2 (A) was measured after 24h, B cell proliferation (B) after 68h. The response is plotted against the mean fluorescence intensity of the B cells after staining with fluorescein isothiocyanate-labeled F(ab')$_2$ goat anti-rabbit IgG. F(ab')$_2$ anti-IgM (●), Fab' anti-IgM (▲), F(ab')$_2$ anti-H-2K (○), Fab' anti-H-2K (△).

Even in the absence of cross-linking, it is possible that membrane Ig is functionally specialized for antigen-processing and presentation, and serves more than a purely passive role in binding antigen to the cell surface. We tested this idea by comparing responses induced by rabbit anti-Ig with those to rabbit antibodies against a different integral membrane protein, H-2K (13). When compared with normal rabbit IgG, IgG from an antiserum to H2K was presented with much greater efficiency, about 10,000-fold (13), showing that at least the major portion of the effectiveness of membrane Ig in antigen presentation could be accounted for by passive binding. To facilitate a direct comparison between anti-H-2K and anti-Ig, a different experimental protocol was used. Small B cells were incubated at 4° in azide with various concentrations of anti-Ig or anti-H-2K and then washed free of unbound antibody. A portion of each sample was stained with goat anti-rabbit Ig, and the relative amount of rabbit Ig on the B cell surface was determined by flow cytometry. The remainder of each sample was put into culture with T cells, and the T cell and B cell responses were plotted as a function of the amount of rabbit Ig on the surface of the B cells at the beginning of culture (Figure 4). As before, Fab$_2$ and Fab anti-Ig gave the same responses from both T cells and B cells, but Fab$_2$ anti-H2 was presented 3- to 5-fold less efficiently than anti-Ig. Although this difference seems small in comparison to the 10,000-fold differences between either antibody and normal rabbit IgG, it was a consistent finding of many experiments with three different T cell lines. It does not represent a simple consequence of fine-specificity of the T cell line, since the line cannot distinguish pre-immune IgG from the rabbits making the two antibodies (13). Nor is it a matter of affinity, since it is highly unlikely that the divalent anti-H-2 has a lower affinity than the monovalent Fab anti-Ig. We do not know whether reduced affinity is involved in the considerably reduced effectiveness of monovalent Fab anti-H-2.

The results shown in Figure 3 and Figure 4 present a paradox, since monovalent and divalent anti-Ig are handled so differently by the B cell. The divalent antibody is cleared from the cell surface in a matter of minutes, while the monovalent anti-Ig persists on the cell surface for hours (16). Moreover, although Fab anti-Ig is, in fact, internalized (much more slowly than Fab2 anti-Ig), class I molecules like H-2K are the most difficult of all membrane molecules to modulate, and endocytosis cannot be measured even after cross-linking (B. Pernis, personal communication). If one believes that antigen presentation by small B cells involves internalization and processing, then it appears that internalization for processing is a subtle process not directly related to gross internalization. The possibility that our T cell lines, unlike most others, see intact antigen on the cell surface is hard to reconcile with presentation of divalent anti-Ig after incubation for 2h at 37° before addition of T cells (unpublished experiment). We are currently exploring the relationship between endocytosis and processing for presentation in our system.

ROLE OF T CELL HELP

Now we turn to the question of how the T cell activates the B cell when it recognizes antigen on the B cell surface. Some form of help other than, or in addition to, the production of the stable helper factors which accumulate in culture supernatants of activated T cells seems to be required since, with one possible exception (17), such culture supernatants do not induce DNA synthesis in small resting B cells without mitogenic concentrations of anti-Ig or some other initial activating signal (18). Also, help can be delivered locally, since when help depends on antigen presentation by B cells, the presenting B cell can receive help under conditions, particularly limiting antigen

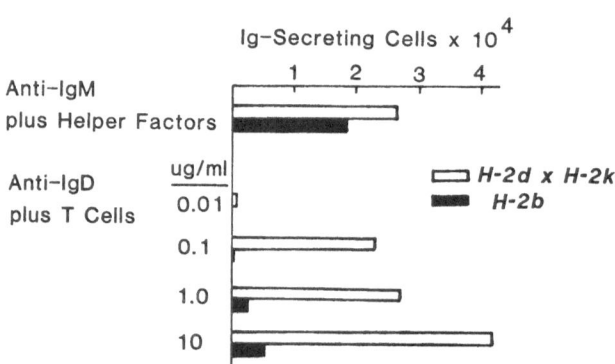

Figure 5. MHC-restriction of the effector phase of T cell help. Small B cells from C3D2 (H-2^k x H-2^d) and B10 (H-2^b) mice were mixed and cultured 4d with 3 x 10^4 T cells (line CDB41) or helper factors (50% mouse splenic Con A supernatant). The T cell line is restricted to H-2^k, and cannot recognize anti-Ig on H-2^b B cells. At the end of the culture period, the H-2 type of the IgM secreting cells was determined by treatment with anti-H-2^k plus complement before the plaque assay. This figure shows part of the experiment from Table III of reference #1.

concentration, when other, "bystander" B cells in the same culture vessel, which cannot act as APC because they bear the wrong MHC allele, do not respond (19,1). Figure 5 shows such an experiment. The secretory response at low concentrations of anti-IgD is derived almost entirely from the presenting B cells. However, at higher concentrations of anti-IgD, we see a small but significant response in the bystander cells. The bystander response is more marked if one looks at B cell proliferation rather than Ig secretion. It does not depend on cross-linking of membrane Ig, since in other experiments (not shown) in which rabbit anti-hapten antibody is attached to the B cells by a hapten sandwich with a haptenated, allotype-specific anti-IgD monoclonal antibody, equivalent bystander responses are obtained from allotype congenic mice which cannot bind anti-Ig.

The leakage of help to bystander B cells implies that MHC-restricted recognition of antigen on the B cell surface may not be part of the delivery of help, i.e., the machinery of recognition may be distinct from the machinery for the delivery of help. In further experiments (not shown), we found that antibodies against both I-A and I-E class II antigens of the bystander cells have absolutely no effect on the response of bystander cells to T cells activated by mitomycin C-treated presenting B cells of a different haplotype. Also, we and other laboratories have been unable to confirm reports of direct stimulatory (20) or inhibitory (21) effects of anti-class II antibodies on B cell responses. Therefore, help does not appear to involve signal transduction by the class II molecule on the B cell. Instead, it seems most likely to involve interaction of other membrane molecules or local action of secreted lymphokines, which may be labile or secreted in a directed manner toward the presenting B cell, as proposed by Janeway elsewhere in this volume.

SMALL B CELLS AS APC

In contrast to our findings, experiments in other laboratories suggest that small, resting B cells are deficient in their ability to present alloantigens or protein antigens to T cells, even when their radiosensitivity is taken into account (22,23). It should be pointed out that experiments reported here were undertaken with T cell lines and hybridomas, which may have less stringent activation requirements than normal or in vivo primed T cells. Normal resting T cells may require an interaction with a specialized accessory APC before they can interact directly with B cells as APC (24). We have not been able to show a requirement for accessory cells in our system. It is harder to explain why our lines appear to differ from other continuous in vitro lines and hybridomas (23, and Grey, this volume) in their ability to see antigen on selected small B cells.

Part of the problem may be inefficient non-specific uptake of antigen by small B cells, since in most of our experiments, presentation by small B cells involves reagents which bind to the B cell surface. However, normal rabbit Ig or antinapten antibodies are also presented to our T cell lines and hybridomas with about equal efficiency by small B cells and anti-Thy1-treated spleen cells, which include large B cells, macrophages, and dendritic cells. Possible differences between resting B cells selected by density on Percoll gradients versus size by centrifugal elutriation need to be explored. Also, it is possible that our T cell lines activate B cells before antigen recognition to convert them into APC. In an attempt to test this possibility, we found that pre-incubation of T cells and small B cells in the absence of antigen did not change either the dose response or the kinetics of the IL-2 response following antigen (not shown). Therefore, if antigen-independent activation of small B cells to APC-status occurs, it must occur more rapidly than other indications of activation, such as increased class II expression, radioresistance, or blast transformation.

A more interesting possibility is that T cells are heterogeneous in their ability to see antigen on small B cells, perhaps because of a difference in post-translational modification of class II molecules on resting versus activated B cells, as proposed by Cowing (23). Although induced and maintained by conventional procedures using normal Ig as antigen, our T cell lines and hybridomas were selected at limiting dilution for their ability to induce B cell responses with anti-Ig, and may be unusually effective at recognizing antigen on small B cells.

REFERENCES

1. Tony, H.-P., and D.C. Parker. 1985. J. Exp. Med. 161: 223.
2. Chesnut, R.W., and H.M. Grey. 1981. J. Immunol. 126: 1075.
3. Cambier, J.C., J.G. Monroe, K.M. Coggeshall, and J.T. Ransom. 1985. Immunology Today 6: 218.
4. Parker, D.C. 1980. Immunol. Rev. 52: 115.
5. Mond,J.J., E. Seghal, J. Kung, and F.D. Finkelman. 1981. J.Immunol. 127: 881.
6. Kurt-Jones, E.A., J.-M. Kiely, and E.R. Unanue. 1985. J. Immunol. 135: 1548.
7. Augustin, A., and A. Coutinho. 1980. J. Exp. Med. 151: 587.
8. Jones, B., and C.A. Janeway, Jr. 1981. Nature (Lond.) 292: 547.
9. DeFranco, A.L., J.D. Ashwell, R.H. Schwartz, and W.E. Paul. 1984. J. Exp. Med. 159: 861.
10. Rock, K.L., B. Benacerraf, and A.K. Abbas. 1984. J. Exp. Med. 160: 1102.
11. Malynn, B.A., and H.H. Wortis. 1984. J. Immunol. 132: 2253.
12. Lanzavecchia, A. 1985. Nature 314: 537.
13. Tony, H.-P., N.E. Phillips, and D.C. Parker. 1985. J. Exp. Med. 162: 1695.
14. Julius, M.H., H. von Boehmer, and C.L. Sidman. 1982. Proc. Natl. Acad. Sci. U.S.A. 79: 1989.
15. Zubler, R.H., and O. Kanagawa. 1982. J. Exp. Med. 156: 415.
16. Schreiner, G.F., and E.R. Unanue. 1976. Adv. Immunol. 24: 38.
17. Leclercq, L., G. Bismuth, and J. Theze. 1984. Proc. Natl. Acad. Sci. U.S.A. 81: 6491.
18. Melchers, F., and J. Andersson. 1985. Ann. Rev. Immunol. 4. In press.
19. DeFranco, A.L., J.D. Ashwell, R.H. Schwartz, and W.E. Paul. 1984. J. Exp. Med. 159: 861.
20. Palacios, R.C., O. Martinez-Maza, and K. Guy. 1983. Proc. Natl. Acad. Sci. USA 80: 3456.20.
21. Forsgren, S., G. Pobor, A. Coutinho, and M. Pierres. 1984. J. Immunol. 133: 2104.
22. Frohman, M., and C. Cowing. 1985. J. Immunol. 134: 2269.
23. Krieger, J.I., S.F. Grammer, H.M. Grey, and R.W. Chesnut. 1985. J. Immunol. 135: 2937.
24. Inaba, K. and R.M. Steinman. 1984. J. Exp. Med. 160: 1717.

CLONAL ANALYSIS OF HUMAN T LYMPHOCYTES

INDUCING B CELL GROWTH

Maria Cristina Mingari and Lorenzo Moretta

Istituto Nazionale per la Ricerca sul Cancro
and   University of Genova
16132 Genova,Italy

INTRODUCTION

The central role of T lymphocytes in the control of antibody responses has been known since long time. Nevertheless,
only during the last few years major progresses have been made towards 1) the understanding of the functional and phenotypic characteristics of T cells which influence B cell growth
and differentiation. 2) The nature of soluble factors involved
in signalling between T and B lymphocytes. 3) The stages and
mechanisms involved in B cell activation, proliferation and
differentiation (1). For the most part, this favorable situation can be traced to the merging of new technologies. On the
one hand, the use of monoclonal antibodies has provided omo-
geneous reagents for the unambiguous definition of the phenotypic characteristics of T cell subsets and clones (2) and of
B cells at different stages of activation. In addition, mAbs
specific for some lymphokines and/or their cell surface receptors   allowed   a more precise definition of the involvement
of these mediators in B cell growth or differentiation (3-6).
At the same time, molecular ingeneering techniques provided
scientists with some lymphokines in recombinant form (7) : it
thus became evident that several different activities could
be attributed to single lymphokines. In addition, rapid advances
in flow microfluorometry have provided a precise and objecti-
ve means to quantify given surface markers and the extent to
which the marker is expressed. Finally the improvement of T
cell cloning techniques made it possible to grow virtually all
T lymphocytes, thus allowing to identify both frequencies and
subset distribution of T cells responsible for B cell proliferation and differentiation.

ASSAYS FOR EVALUATION OF B CELL PROLIFERATION

Sensitive assays have been developed in recent years for ana-

lysis of activation, proliferation and differentiation of B lymphocytes. Two assay systems have been applied in humans for analysis of T cell-promoted B cell proliferation. Both of these systems act by polyclonally triggering human B cells via surface Ig. The first system is based on the use of Staphilococcus aureus Cowans strain I.
(SAC) (8), which binds to both FC and Fab portions of human Ig. SAC delivers a strong activation signal to the B cells via the crosslinking of surface Ig. It is noteworthy that SAC, in addition to B cell activation, also drives B cell through cell proliferation, even in the absence of BCGF. SAC-induced B cell proliferation peaks at day 3 and sharply declines by day 5 and 6. The principle of the assay is to preactivate B cells with SAC for 3 days prior to addition to the culture of supernatant containing BCGF activity. Under these conditions, B cell proliferation is normally detected after an additional two to three days of culture (8). The other assay model for measuring BCGF activity is based on B cell activation by soluble anti-μ antibody and the simoultaneus addition of a source of supernatant to be tested for BCGF activity (9). The major effect of SAC and anti-μ antibody in making B cell responsive to BCGF consists in promoting B cell activation, which can be evaluated by the detection of RNA synthesis de novo and by the increase in cell size. In addition, the activation process is associated with the sequential expression of markers such as the 4F2 antigen and the transferrin receptor.

LYMPHOKINES INDUCING B CELL GROWTH

For several years it has been assumed that BCGF and IL-2 were different molecules and that the latter could only act as a T cell growth factor. However, when the recombinant (r) IL-2 became available, it became clear that IL-2 could act as a potent BCFG both in mouse (10) and in man (4). Thus, we showed that r-IL-2 derived from Escherichia coli expressing the human gene as well as IL-2 obtained by affinity chromatography using specific mAbs promoted strong proliferation of SAC-activated human B cells (11). Moreover, we demonstrated that the anti-TAC mAb (reacting with the IL-2 receptor molecules expressed by activated T cells) also reacts with SAC-activated B cells and sharply inhibited the proliferative response of such cells to IL-2. Finally, immunoprecipitation experiments revealed that anti-TAC mAb defines molecules of identical molecular weight on activated T and B cells (4).
Although BCGF activity measured in the SAC system may be largely (if not totally) related to IL-2, several experimental evidences indicated that molecules distinct from IL-2 may act as as BCGF (12,13). Along this line, Romagnani and colleagues inequivocally showed that another well defined lymphokine in recombinant form (r -interferon) promoted proliferation of anti-μ activated human B lymphocytes (14). γ-IFN was found to be responsible for part of the BCGF activity contained in polyclonal

T cell-derived SN (measured in the anti-u proliferation assay).
This was shown by the use of mAbs specific for human γ-IFN :
addition of such mAb resulted in the abrogation of the BCGF
activity of rγ-IFN and in the partial inhibition of the BCGF
activity of spleen or tonsil-derived polyclonal crude SN (14).
Interestingly, the combined use of anti-TAC and antiγ-IFN mAbs
resulted in a strong inhibition of B cell proliferation promoted by polyclonal SN. Analysis of the relative role of γ-IFN
and IL-2 in inducing proliferation of anti-μ activated B cells
provided evidence in favour of a synergistic effect of these
lymphokines. Thus, even low concentrations of γ-IFN strongly
enhanced the B cell responses to r-IL-2 : in addition, in all
istances BCGF activity induced by the two lymphokines used in
combination was greater than the sum of the activities induced
by the lymphokines used alone. Since other investigators provided evidence that γ-IFN plays a role in the differentiation of
murine and human B cells (15,14), it appears important to define whether γ-IFN acts on B cells at different stages of activation and proliferation or rather whether the γ-IFN-induced
Ig production represents a direct consequence of the γ-IFN-sti-
mulated B cell proliferation. The fact that IL-2 and γ-IFN act
in a sinergistic fashion on B cell proliferation suggests that
the two lymphokines interact with different cell surface receptors. Very recent studies suggested that γ-IFN may be responsible for the recruitement of a larger number of IL-2 receptor-
positive B lymphocytes (17).

SURFACE PHENOTYPE AND FUNCTION OF T CELLS RESPONSIBLE FOR
BCGF ACTIVITY

It is common notion that human T cells able to promote B cell
proliferation belong to the so called "helper/inducer" subset
expressing the T4 surface antigen (2). However, recent studies
at the clonal level clearly indicated that T cell clones producing IL-2 or displaying BCGF activity may belong to the T8[+]
subset (18). In order to provide a precise quantitative assessment of the pool size and subset distribution of T cells capable of inducing B cell growth, we applied a microculture system
which allows clonal proliferation and expansion of virtually
100% peripheral blood resting T lymphocytes (19). Clones so derived were analyzed for their ability to produce BCGF in either
the SAC or the anti-μ driven B cell proliferation assay.
The experimental design consisted in cloning FACS-purified T4[+]
and T4-(T8+) T cells under limiting conditions with 10[5] spleen
cells as feeder cells and 1% PHA. IL-2-containing spleen SN was
added after 24 h. Growing microcultures were subsequently
screened for their ability to release BCGF upon stimulation
with PHA. Since in these experiments the cloning efficiency was
96%, the set of clones derived could be considered to be largely representative of the original populations. Microcultures
obtained by plating T cells at 0.3 cells/well were operationally
considered as clones.. It was found that out of 100 T8[+] clones

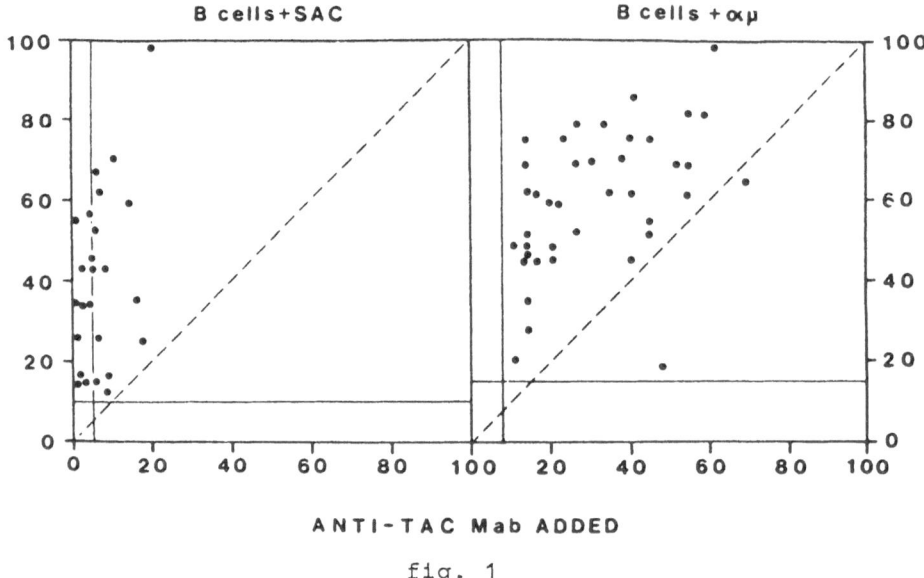

fig. 1

Effect of a mAb specific for the IL-2 receptor on B cell proliferation induced by BCGF-producing T cell clones. T cell clones were derived from peripheral blood by applying a culture
system which allows clonal proliferation of virtually 100% T
lymphocytes. SN of PHA-stimulated clones were screened in the
SAC-driven prestimulation assay or in the anti-μ costimulation
assay. SN of clones displaying BCGF activity were analyzed
again for their ability to induce B cell proliferation in the
presence or in the absence of the CM269 (anti IL-2 R) mAb. Data are expressed as a percentage of the B cell proliferation
induced by the clone with the maximal activity in each experimental group. Note that the CM269 mAb abrogates B cell proliferation induced by all clones in the SAC system, but not in
the anti-μ assay.

screened, 13 had BCGF activity in the SAC assay and 15 in the
anti-u assay.
Among T4[+] clones, 74% displayed BCGF activity in the SAC assay
and 79% in the anti-μ assay. These data provide minimal esti-
mates of the percentages of T cells capable of inducing B cell
growth. They also confirm the concept that T cells capable of
inducing B cell proliferation in the SAC or anti-μ driven assay
are not confined to the T4[+] subset (18). In this context, Moretta showed that IL-2 producing cells represented 70-90% of PB

$T4^+$ cells, whereas they represented 10-15% of $T8^+$ cells (20). Since the SAC-driven BCGF assay appears to detect mostly IL-2, it is not surprising that the figures obtained were similar to those of a typical IL-2 assay (such as the CTLL proliferation assay) it is noteworthy that a remarkable proportion of $T8^+$ cells had BCGF activity. $T8^+$ cells are known to display cytolytic activity, moreover, by using a high cloning efficiency culture system Moretta and coll. showed that virtually 100% $T8^+$ cells give rise to a cytolytic progeny (21). Therefore, it is not unusual to detect cytolytic T cells which may induce B cell growth as well.

Given the finding that virtually all the BCGF activity of a polyclonal SN in the SAC-driven assay could be inhibited by the addition of mAbs directed to the IL-2 receptor, it was of interest to analyze the effect of such mAb on the BCGF activity of clonal SN active in the SAC or in the anti-u assay. As shown in fig. 1, a strong inhibition of B cell proliferation was observed in the SAC assay for all BCGF-producing clones. On the contrary, in the anti-u assay, the SN of several clones were clearly resistant to, or partially inhibited by the anti-IL-2 receptor mAb.

In view of the demonstration that -IFN can act as BCGF in the anti-u assay it was of interest to analyze the SN of clones resistant to anti-IL-2 receptor mAb for both IL-2 and -IFN activity. None of the 9 clones analyzed (derived from two different donors) had IL-2 activity, whereas 6 had -IFN activity. These experiments again indicate that the SAC system is primarily susceptible to IL-2, whereas the anti-u system responds to IL-2 and to other BCGF (s).

REFERENCES

1)-Moller G. (Ed.), Immunol. Rev. 78,1984.

2)-Reinherz, E.L., Schlossman, S.F. cell, 19, 821, 1980.

3)-Tsudo, M., Uchiyama, T., Huchino, H. J. Exp. Med. 160, 612, 1984

4)-Mingari, M.C., Gerosa, F., Carra, G., Accolla, R.S., Moretta, A., Zubler, R.H.,Waldmann, T.A., Moretta, L. Nature 312, 641, 1984

5)- Waldmann, T.A., C.K. Goldman, R.J. Robb, J.M. Depper, W.J. Leonard, S.O. Sharrow, K.F. Bongiovanni, S.J. Korsmeyer,and W.C. Green., J. Exp. Med. 160, 1450, 1984.

6)- Muraguchi, A., J.H. Kehrl, D.L. Longo, D.J. Volkman, K.A. Smith, and A.S. Fauci. J. Exp. Med. 161, 181, 1985

7)-Devos, R. Plaetnick, G., Cherautre, H., Simons, G., Degrave, W., Tavernier, J., Rement, E. and Fiers, W. Nucleric Acid Res. 11, 4307, 1983.

8)-Romagnani, S., M.G. Giudizi, R. Biagiotti, F. Almerigogna, E. Maggi, G.F. Del Prete, and M. Ricci. 1981. Surface immuno globulins are involved in the interaction of protein A with human B cells and in the triggering of B cell proliferation induced by protein A-containing Staphylococcus aureus. J. Immunol. 127:1307.

9)-Muraguchi, A., and A.S. Fauci. J. Immunol. 129:1104, 1982.

10)-Zubler, R.H., Lowenthal, J.W., Erard, F., Hashimoto, N., Devos, R., Mac Donald H.R.: J. Exp. Med., 160, 1170, 1984.

11)-Mingari, M.C., Gerosa, G., Moretta, A., Zubler, R.H., Moretta, L. Eur. J. Immunol., 15, 193, 1985.

12)-Butler, J.L., A. Muraguchi,H.C. Lane, and A.S. Fauci. J. Exp. Med. 157, 60, 1982.

13)-Okada, M, N. Sakaguchi, N. Yoshimura, H. Hara, K. Shimuzu, N. Yoshida, K. Yoshizaki, S. Kishimoto, Y. Yamamura, and T. Kishimoto. J. Exp. Med. 157. 588, 1983.

14)-Romagnani, S., Giudizi, M.G., Biagiotti, R., Almerigogna, F., Mingari, M.C., Maggi, E., Liang, C-M., Moretta, L. J. Immunol. 1986 (in press).

15)-Sidman, C.L., J.D. Marshall, L.D. Shultz, P.W. Grey, and H.M. Johnson. Nature (Lond.). 309, 801, 1984.

16)-Nakagawa, T., T. Hirano, N. Nakagawa, K. Yoshizaki, and T. Kishimoto. J. Immunol. 134, 959, 1985.

17)-Romagnani, S., Giudizi, M.G., Almerigogna, F., Biagiotti, R., Alessi, A., Mingari, M.C., Liang, C-M, Moretta, L., Ricci, M. Eur. J. Immunol. 1986. (in press).

18)-Mingari, M.C., Moretta, A., Maggi, E., Pantaleo, G., Gerosa, F., Romagnani, S., Moretta, L. Eur. J. Immunol. 14, 1066.

19)-Moretta, A., Pantaleo, G., Moretta, L., Mingari, M.C., Cerottini, J.C. Direct demonstration of the clonogenic potential of every human peripheral blood T cells. Clonal analysis of HLA-DR expression and cytolytic activity. J. Exp. Med. 157, 743, 1985.

20)-Moretta, A., Frequency and surface phenotype of human T lymphocytes producing Interleukin-2. Analysis by limiting dilution and cell cloning Eut. J. Immunol. 15, 148, 1985

21)-Moretta, A., Pantaleo, G., Moretta, L. Mingari, M.C., Cerottini, J.C. J. Exp. Med. 158, 571, 1983.

PRIVATE AND SHARED IDIOTYPIC DETERMINANTS OF THE

HUMAN T CELL ANTIGEN RECEPTOR

Robert D. Bigler and Nicholas Chiorazzi

The Rockefeller University

New York, New York

INTRODUCTION

T lymphocytes play a crucial role in initiating most immune responses. Not only are T cells, especially antigen-specific helper T cells, necessary to induce functional effector T cell subpopulations but also to initiate development of the B cell response. The complete understanding of this co-ordinated T-B response requires knowledge of both the mechanism of T cell activation and the method by which T cells transmit signals to B cells. Once activated, T lymphocytes appear to function by secreting various soluble factors. These factors have been reported to influence B cell proliferation, differentiation, and immunoglobulin isotype selection (1-3). The nature of the initial events in triggering T lymphocytes to progress from a resting to activated state is poorly understood. Understanding the requirements to initiate this specific T cell activation, however, is crucial to understanding the integration of these two major functional arms of the immune system. The recent definition of the T cell antigen receptor using both monoclonal antibodies (mAbs) and cDNA probes has permitted a new, more precise method for investigating both the cellular events and molecular structures involved in this initial stage of T-B interaction.

DEFINITION OF THE T CELL ANTIGEN RECEPTOR BY MONOCLONAL ANTIBODIES

Prior to the production of mAbs, several groups attempted to define the structure of the T cell antigen receptor by raising antisera against T cell specific surface antigens, immunoglobulin variable regions, or soluble T cell factors (4-10). These antisera were used to search for antibodies that bound specifically to T cells, that inhibited a specific function, and that would immunoprecipitate the reactive molecule detected in these systems (11,12). These studies, however, were frought with many technical difficulties due to the necessity of purifying polyclonal antisera to define an individual specificity on the T cell surface. The development of monoclonal antibody technology has permitted a more reproducible means of defining this molecule.

The first mAb reacting with a putative T cell antigen receptor molecule was reported by Allison et al. (13). Using a mouse T cell line as the immunizing T cell, a mAb was produced that reacted exclusively

TABLE I

SURVEY OF THE REACTIVITY OF TWO ANTI-T CELL
ANTIGEN RECEPTOR ANTIBODIES S160 and S511

	Leukemia SU	Leukemias T cell	Cell Lines T	B	Myeloid	Normal Cells T	non-T	Thymus	Blasts
S511	>85	-	-	-	-	1.6	-	0.7	1.3
S160	>85	-	-	-	-	-	-	-	-

Mean values of the percentage of reactive cells in each cell type; represents <0.5% reactive cells. Data adapted from Bigler et al. 1982 J. Exp. Med.158: 1000.

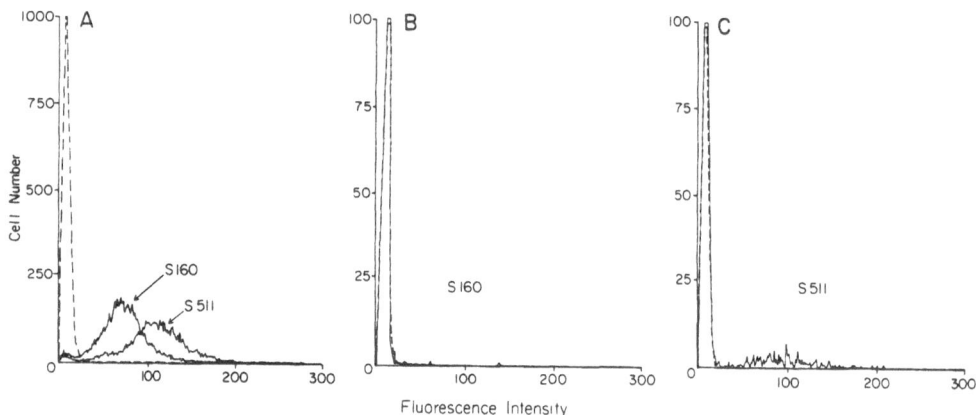

Figure 1. Fluorescence histograms of leukemia SU (A) and normal human T lymphocytes (B and C) stained with mAb S160 or with mAb S511.

with the immunizing cells and not with other T cells or non-T cells. This mAb immunoprecipitated a 75kD disulfide-linked heterodimer composed of a 41kD chain and a 39kD chain. Using diagonal sodium dodecyl sulfate-polyacrylamide gel electrophoresis (SDS-PAGE) analysis this study also indicated that a similar molecule was present on normal T cells. Following this report, studies by Haskins et al.(14), using an antigen specific mouse T-T hybridoma, by Meuer et al.(15), using an antigen specific human cytotoxic T cell clone, and by our lab (16), using a human Sezary cell leukemia, were published defining similar molecules. In our studies, the leukemic T cells used to immunize mice had a surface phenotype of OKT-3[+], OKT-4[+], OKT-8[-], and anti-Tac[-]. Two mAbs, S160 and S511, were described from an initial panel of 610 hybridomas. These two mAbs stained the same percentage of leukemic cells from cryopreserved samples of peripheral blood obtained from the patient SU (Fig. 1-A and Table I). These antibodies did not stain small lymphocytes in the same sample when observed by fluorescence microscopy. Other leukemic T cells were screened by flow cytometry and were non-reactive as were various T cell lines, B cell lines, and myeloid cell lines. When normal peripheral blood mononuclear cells were studied there was a distinct difference between these two mAbs. Mab S160 did not react with any peripheral blood cells, including T cells (Fig. 1-B), while mAb S511 stained a small population of T cells (Fig. 1-C), thymocytes, and mitogen-activated T cell blasts. Thus, of the two mAbs described in the original study, one reacted exclusively with a private idiotype on the leukemic T cells and the other showed a similar restricted specificity for the tumor cells except for cross-reactiing with

1-2% of normal peripheral blood T lymphocytes in all individuals. That these two mAbs were detecting different epitopes on the same molecule was confirmed when each mAb immunoprecipitated the same 80kD disulfide-linked heterodimer from the leukemic cells as demonstrated by one- and two-dimensional SDS-PAGE. The two chains comprising this molecule on the leukemic cell had a mobility of 43kD and 38kD and both possessed a neutral isoelectric point in contrast to the 49kD and 43kD structure with acidic and neutral isoelectric points respectively described for a different human T cell antigen receptor (15). This study thus described a human T cell antigen receptor molecule on leukemic T cells which was similar to other putative receptor molecules reported on murine and human T cell lines and also demonstrated that mAb S511 detected a cross-reactive specificity presumably present on the T cell antigen receptor of normal T cells.

The initial studies of Haskins et al. (14) and Meuer et al. (15) correlated the structure of this molecule with its function by demonstrating that the response of antigen-specific murine T-T hybridomas or human interleukin 2 (IL-2) dependent cytotoxic T cell clones could be specifically blocked by mAbs detecting a private idiotype on their respective cell lines. These studies and others (17,18) also proved that when mAbs were present either complexed to Sepharose or free in solution they could specifically augment the baseline proliferation of these previously activated cell lines. We used a second T cell leukemia, which was OKT-3[+], OKT-4[+], and anti-Tac[+], to produce another mAb with private idiotypic specificity (19). In this study, we compared the ability of two different anti-receptor mAbs to activate the respective leukemic cells. We demonstrated that when the new mAb was complexed to Sepharose, specific proliferation of the leukemic cells could be produced. The SU leukemic cells, in contrast, could not be induced to proliferate when cultured with S160 or S511, even if additional growth factors were provided. This study with leukemic cells expressing different activation antigens complemented the previous studies and confirmed that mAbs reacting with this heterodimer could induce proliferation of the reactive cells. Thus, although mAbs with private idiotypic specificities could block antigen specific function and induce proliferation of previously activated cells, these mAbs had not been able to activate T cells not already demonstrating evidence of prior activation.

DEFINITION OF THE T CELL ANTIGEN RECEPTOR GENES

As these studies using mAbs were being reported, Yanagi et al. (20) and Hedrick et al. (21,22) simultaneously reported the cloning of a T cell specific gene. Using a subtractive hybridization protocol, these two groups cloned a gene which possessed ~30% homology to immunoglobulin, especially to light chains, when the predicted amino acid composition was compared. A more striking relationship to immunoglobulin was evident in the division of these genes into V, D, J, and C regions with intra-chain disulfide bonds in positions almost identical to those present in immunoglobulin V regions. As initially defined, these genes were only T cell specific and were not proven to code for the molecule defined by mAbs. This connection, however, was rapidly provided by Acuto et al. (23) when N-terminal sequencing of the beta chain of a unique human T cell line was shown to be almost identical to the amino acid sequence predicted by the MOLT 3 gene (20).

Using a similar approach to that used for cloning the beta chain, Chien et al. (24) and Saito et al. (25) reported a cDNA clone consistent with the chemical features of the murine alpha chain. Using a human T cell line, HPB-MLT, Hannum et al. (26) used tryptic peptide fragments of the alpha chain to obtain peptide sequences of multiple fragments.

After generating appropriate oligonucleotides based on these sequences, Sim et al. (27) reported the cloning of the human alpha chain gene. The results of these and other studies all indicated that the T cell antigen receptor uses recombination strategies similar to those documented for immunoglobulin (28-32). The association of the T cell antigen receptor genes as members of the immunoglobulin supergene family (33) suggests that the private idiotypic, cross-reactive idiotypic, framework, constant region, and other specificities detected on immunoglobulin might be defined on the T cell antigen receptor molecule as well. This type of structural arrangement of the receptor was predicted by our studies two years ago when we described the private specificity of mAb S160 and the cross-reactive specificity of mAb S511(16). Subsequently, antibodies to constant region (34,35), allotypic (36), and V_{beta} framework (37) epitopes have been described.

ACTIVATION OF NORMAL RESTING T LYMPHOCYTES

The T cell antigen receptor, by definition, must be able to specifically initiate the proliferation of resting T lymphocytes. As mentioned above, all previous studies evaluating the function of this putative T cell antigen receptor molecule utilized cells showing evidence of prior activation. These studies used either blocking of antigen specific responses or augmenting proliferation to support the assumption that this molecule was indeed the T cell antigen receptor. Although these were valid methods to test whether this molecule functioned as expected of the antigen receptor, this approach is not as critical a test as the ability to specifically activate resting cells. We have used the normal T cell reactivity of mAb S511 to demonstrate precisely that function.

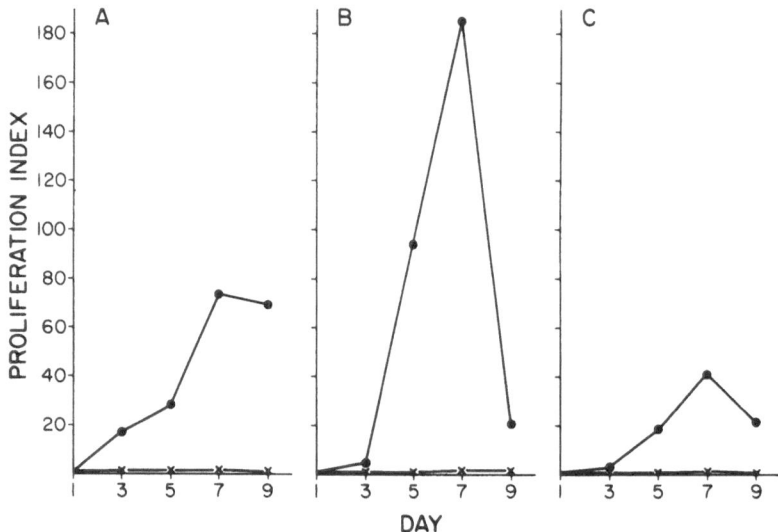

Figure 2. Kinetics of proliferation of normal T cells cultured with irradiated non-T cells and with soluble S511 mAb (●) or soluble anti-DNP mAb (×) at 1.6mg/ml. Proliferation index is a measure of [^{3}H] thymidine incorporation defined as (mean cpm of cells cultured in mAb)/ (mean cpm of cells cultured in media alone). Adapted from Bigler et al. J.Exp.Med 161:1450 (1985).

126

TABLE II
Normal T Cell Proliferation Induced by Monoclonal Antibody S511

	T cells + non-T cells[1]	T cells alone[1]
	Day 7	Day 7
Anti-DNP[2]		
10 µg/ml	1.6	1.5
1	1.2	1.2
0.1	1.8	1.0
S511		
10 µg/ml	61.3	79.9
1	49.3	126.8
0.1	2.0	14.3
IL-2 10%[3]	22.0	ND[4]
PHA 1:100	9.9	20.6

1) Mean proliferation indices as defined in Fig. 2.
2) Anti-DNP mAb was purified using the same technique as mAb S511.
3) IL-2 supplemented media without addition of mAb.
4) ND – Not done.
5) Data adapted from Bigler et al. 1985. J.Exp.Med. 161:1450

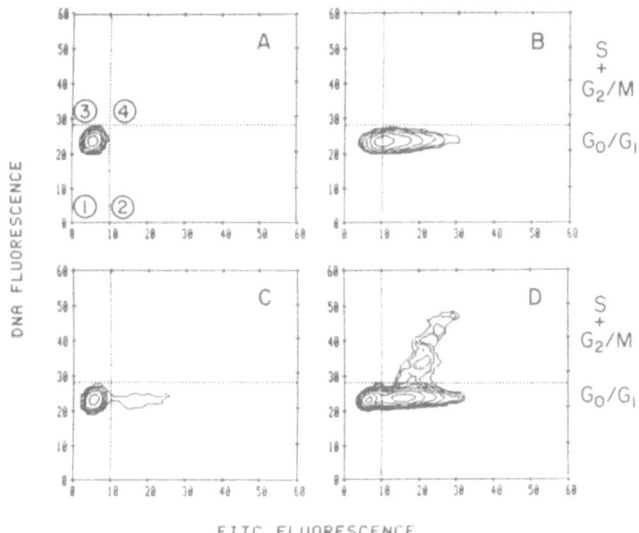

FITC FLUORESCENCE

Figure 3. Simultaneous surface immunoflourescence and cell cycle analysis of normal circulating human T cells. T cells were stained by indirect immunofluorescence using FITC-Goat anti-mouse immunoglobulin alone (A), mAb 9.6 which reacts with the sheep erythrocyte receptor (B), mAb S511 (C), and a cell lined enriched for S511 expression stained with mAb S511 (D). Each sample was then fixed in ethanol and stained with propidium iodide for cell cycle analysis. In each panel the distinction between G_0/G_1 and $S+G_2/M$ and surface fluorescence-positive and -negative is provided. Adapted from Bigler et al. J.Exp.Med. 161:1450 (1985).

Normal T lymphocytes were cultured in the presence of 10% irradiated non-T cells and purified soluble mAb S511. S511 mAb induced proliferation of the T cells in these cultures as measured by thymidine
incorporation (Fig. 2) (38). The proliferative response reached a maximum on day 7 and was dependent on the concentration of the mAb as shown
in Table II. Parallel experiments demonstrated that this T cell proliferation could be induced by mAb S511 without the addition of irradiated
non-T cells (Table II). Although confirming that S511 mAb could induce
proliferation, these studies did not prove that the S511$^+$ cells were the
proliferating cells. To confirm that the cells in this system were
S511$^+$, T cells were stimulated for 7 days with mAb S511, washed, and the
proliferating cells were expanded for two additional days in media containing 10% IL-2. This study demonstrated that in a population shown to
be 88% OKT-3$^+$ and 40% Ia$^+$, 34% of the cells were S511$^+$. Cells grown in
media, PHA, or IL-2 failed to demonstrate an S511-reactive population
greater than ~2%. Thus, the proliferating cells in cultures activated
by S511 mAb were predominantly S511$^+$ cells rather than non-specifically
recruited cells.

This expansion of S511$^+$ cells could have related to prior activation of these cells comparable to the other systems studied. To assess
whether S511$^+$ T cells showed evidence of prior activation, freshly
prepared peripheral blood T cells were stained for S511 expression using
indirect immunofluorescence, fixed with ethanol, and then stained with
propidium iodide. In this way simultaneous cell cycle analysis and
immunofluorescence analysis of the S511-reactive cells could be examined
by flow cytometry. These studies, an example of which is displayed in
Fig. 3, demonstrated that circulating T cells are in the G_0/G_1 phase
(Fig. 3-A and B) and that the S511$^+$ cells are also in this phase (Fig.
3-C). This evidence, plus the lack of preferential expansion of S511$^+$
cells in IL-2 media, demonstrated that S511$^+$ cells are in the G_0 or G_1
phase of the cell cycle which is consistent with the definition of a
resting T cell.

TABLE III

PERCENTAGE OF CELLS BEARING OKT-4 AND OKT-8
ON S511$^+$ T CELL LINES[1]

T cell line[2]	S511	OKT4	OKT8
1	85	69	17
2	61	55	12
3	86	4	63
4	60	2	35
5	82	32	77
6	93	47	90
7	83	6	3

1) Percentage of reactive cells in uncloned cell line measured by indirect
   immunoflurescence and flow cytometry.
2) S511$^+$ cell lines expanded in IL-2 supplemented media using periodic
   addition of S511-Sepharose

In addition to activating S511[+] cells with soluble mAb, we have also used S511-Sepharose to induce proliferation of these cells. Peripheral blood T cells were cultured with S511-Sepharose in IL-2 containing media. Cells would proliferate under these conditions and were expanded in this media. These proliferating cells could be maintained for several months with periodic addition of S511-Sepharose and IL-2. When analyzed by immunofluorescence these cell lines were usually 80-95% S511[+]. To determine the phenotype of these S511[+] cells, we used immunofluorescence and mAbs OKT-4 and OKT-8. When several cells lines were stained, they appeared to be polyclonal since different phenotypes were demonstrated as previously reported (38). These lines could be composed of S511[+] cells expressing primarily OKT-4 or OKT-8 (Table III, lines 1-4), possible expressing both OKT-4 and OKT-8 (lines 5 and 6), and expressing neither OKT-4 nor OKT-8 (line 7). Similar observation have been reported by other investigators with anti-T cell antigen receptor mAbs demonstrating cross-reactivity to normal cells (37,39).

These studies have permitted us to define not only a private idiotype on the antigen receptor of the immunizing leukemic T cells but also a cross-reactive specificity on the antigen receptor of normal T cells. We have demonstrated that these normal T cells bearing the S511 epitope are resting cells that can be preferentially activated by mAb S511. Thus, the S511 mAb replaced processed antigen as one of the initial triggering signals to resting T cells providing strong evidence that the molecules described above are indeed the T cell antigen receptor. Using this approach, it should be possible to dissect the minimal steps required for T cell activation and proliferation and thereby more fully understand the nature of the integration of T-B interaction.

REFERENCES

1. Howard, M., & Paul, W.E. Ann. Rev. Immunol. 1:307-333 (1983).
2. Mayer, L.M., Fu, S.M., & Kunkel, H.G. Immunol. Rev. 78:119-135 (1984).
3. Marrack, P. et al. Immunol. Rev. 63:33-49 (1982).
4. Binz, H., & Wigzell, H. J. Exp. Med. 142:197-211 (1979).
5. Krammer, P.H., Rehberger, R., & Eichmann, K. J. Exp. Med. 151:1166-1182 (1980).
6. Rubin, B., Hertel-Wulff, B., & Kimura, A. J. Exp. Med. 150:307-321 (1979).
7. Lea, T. et al. J. Immunol. 122:2413-2417 (1979).
8. Eichmann, K. Adv. Immunol. 26:195-254 (1977).
9. Tokuhisa, T., & Taniguchi, M. J. Exp. Med. 155:126-139 (1982).
10. Nagy, Z.A. et al. Eur. J. Immunol. 12:393-400 (1982).
11. Binz, H., & Wigzell, H. J. Exp. Med. 154:1261-1278 (1981).
12. Rosenstein, R.W. et al. Proc. Natl. Acad. Sci. USA 78:5821-5825 (1981).
13. Allison, J.P., McIntyre, B.W., & Bloch, D. J. Immunol. 129:2293-2300 (1982).
14. Haskins, K. et al. J. Exp. Med. 157:1149-1169 (1983).
15. Meuer, S.C. et al. J. Exp. Med. 157:705-719 (1983).
16. Bigler, R.D. et al. J. Exp. Med. 158:1000-1005 (1983).
17. Kappler, J.R. et al. Cell 34:727-737 (1983).
18. Meuer, S.C. et al. Science 222:1239-1241 (1983).
19. Posnett, D.N. et al. J. Exp. Med. 160:494-505 (1984).
20. Yanagi, Y.Y. et al. Nature (London) 308:145-149 (1984).
21. Hedrick, S. et al. Nature (London) 308:149-153 (1984).
22. Hedrick, S. et al. Nature (London) 308:153-158 (1984).
23. Acuto, O. et al. Proc. Natl. Acad. Sci. USA 81:3851-3855 (1984).
24. Chien, Y. et al. Nature (London) 312:33-35 (1984).
25. Saito, H. et al. Nature (London) 312:36-40 (1984).

26. Hannum, C.H. et al. Nature (London) 312:65-67 (1984).
27. Sim, G.K. et al. Nature (London) 312:771-775 (1984).
28. Chien, Y. et al. Nature (London) 309:322-326 (1984).
29. Siu, G. et al. Cell 37:393-401 (1984).
30. Malissen, M. et al. Cell 37:1101-1110 (1984).
31. Sims, J.E. et al. Nature (London) 312:541-545 (1984).
32. Duby, A.D. et al. Science 228:1204-1206 (1985).
33. Hood, L., Kronenberg, M., & Hunkapiller, T. Cell 40:225-229 (1985).
34. McIntyre, B.W. & Allison, J.P. Cell 34:739-746 (1983).
35. Brenner, M.B. et al. J. Exp. Med. 160:541-551 (1984).
36. Haskins, K. et al. J. Exp. Med. 160:452-471 (1984).
37. Acuto, O. et al. J. Exp. Med. 161:1326-1343 (1985).
38. Bigler, R.D., Posnett, D.N., & Chiorazzi, N. J. Exp. Med. 161:1450-1463 (1985).
39. Moretta, A. et al. J. Exp. Med. 162:1393-1398 (1985).

MOLECULAR MARKERS OF CLONALITY, LINEAGE, DIFFERENTIATION, AND TRANSLOCATION IN B CELL NEOPLASMS

Stanley J. Korsmeyer, Ajay Bakhshi, John J. Wright,
Winfried C. Graninger, Carolyn Felix, and Masao Seto

Metabolism Branch, National Cancer Institute, National
Institutes of Health, Bethesda, Maryland USA 20892

The demonstration that distinct types of lymphoid neoplasms could be assigned to stages of B or T cell development provided great insights into the biology of these malignancies[1,2]. Historically this has been approached by utilizing cell surface markers associated with various developmental stages of B or T cell maturation. Despite a large number of lineage-associated cell surface markers, it is still frequently impossible to conclusively classify a lymphoid neoplasm as B or T cell in origin. This is often due to the admixture of large numbers of nonneoplastic cells with the neoplastic cells in a lymphomatous tissue. Alternatively, other malignancies may be at a stage of differentiation prior to the expression of any lineage-restricted cell surface antigen. Moreover, the determination of clonality in lymphoid neoplasms has for all practical purposes been limited to the mature B cell malignancies that display the presence of but one light (L)-chain isotype, κ or λ. The DNA rearrangements which assemble the gene subsegments for antigen-specific receptors in B cells as well as T cells serve as molecular markers which are unique to individual neoplasms. The rearranging determinants have made an enormous contribution to refining the diagnosis and understanding the pathogenesis of lymphoid neoplasms.

The rearrangements of immunoglobulin (Ig) genes and the β chain of the T cell antigen-specific receptor (TCR) have proven not only to be lineage associated, but are also extremely sensitive markers for clonality[3,4]. They have resolved issues concerning the cellular lineage commitment, clonality, and stage of differentiation of lymphoid neoplasms that were uncertain by phenotypic analysis alone. In addition, further recombinations of Ig and TCR loci have been defined which contribute directly to the malignant phenotype. This latter type of rearrangement occurs between nonhomologous chromosomes and translocates a cellular oncogene into an Ig or TCR gene locus and its influence[5-7].

DNA REARRANGEMENTS OF Ig GENES GENERATE UNIQUE CLONAL MARKERS

The mechanism which activates an Ig gene generates a clonal marker specific to that particular B cell. This is the result of the process of DNA rearrangement which moves and combines the separated gene segments which encode the final Ig molecule. The act of V/J joining introduces a new restriction endonuclease site and alters the size of the restriction fragment containing the κ gene.

Therefore, a rearranged Ig gene is found on a different sized DNA restriction fragment as compared to its germline or embryonic form. A polyclonal population of normal B cells is comprised of many different cells which possess numerous different Ig gene rearrangements; none of the rearranged genes in this collective population will be discernible by Southern blot analysis because they fall below the threshhold of detection[8]. In contrast, a monoclonal expansion of B cells represents the progeny of an original single cell and possesses multiple copies of the same DNA rearrangement unique to that cell. Such clonal DNA rearrangements are detectable even if the clonal B cell represents only 1-5% of the total cells present[3,9]. Thus lineage-associated DNA rearrangements serve as sensitive as well as specific clonal markers.

Ig GENE PATTERNS IN MATURE B CELLS AND NON-B CELLS

Mature B cells must possess a rearranged heavy (H)- and L-chain Ig gene responsible for their cell surface Ig. Roughly 60% of B cell neoplasms produce κ L-chain, while 40% produce λ L-chain. Of note, κ-producing B cells displayed at least one κ gene rearrangement, while they retained their λ genes in the germline form (Fig. 1). In striking contrast, λ-producing B cells, while displaying the obligate λ rearrangement, had surprisingly rearranged or usually deleted their κ genes[10]. This proved to result from an ordered sequence of L-chain gene events in man in which κ rearrangements preceded λ (Fig. 2)[8,11].

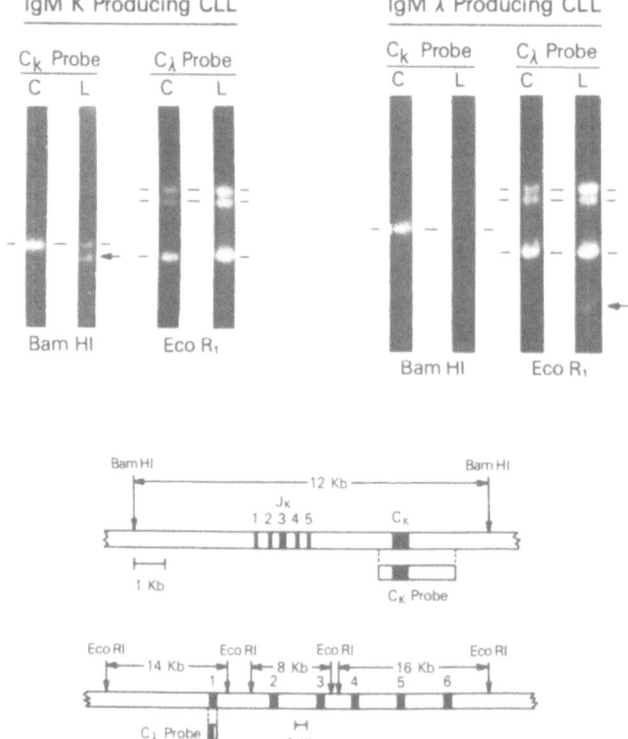

Fig. 1. L-chain genes in κ- and λ-producing B cell chronic lymphocytic leukemia (L) as compared to their fibroblasts (C). A rearranged κ gene (arrow) is present in the κ-CLL while its λ genes are in the germline form (dash marks). In contrast the λ-CLL reveals a loss of κ genes. Lower schematics reveal probes and restriction maps.

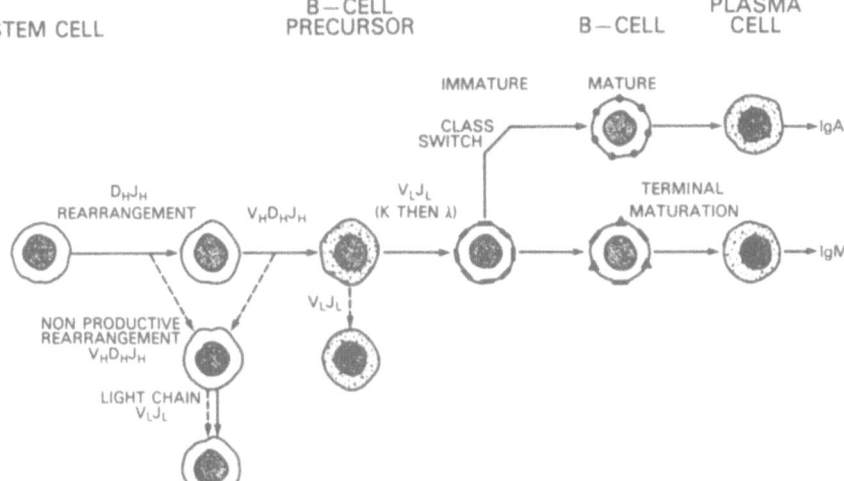

Fig. 2. B cell precursor leukemias represent developmental stages in which
rearrangement of the H-chain D_H segment to J_H segment is generally
followed by the addition of a V_H region. Subsequently L-chains
rearrange in a κ before λ fashion. Dashed lines indicate cells
possessing nonproductive, aberrantly rearranged genes.

Central to this sequential use of L-chain genes is an unanticipated
deletion of the constant κ (C_κ) gene and κ enhancer sequence that precedes
λ rearrangement. All deletions of κ genes are mediated by a uniform κ:de-
leting element (κde)[12]. The κde recombines site specifically with a palindromic signal (CACAGTG) located in the J_κ-C_κ intron in 75% of the instances
of κ gene loss. In the remaining 25% of cases of κ gene loss the element
rearranges with upstream sequences, presumably the heptamer signal of V_κ
regions. In contrast, the κde remained in its germline form on all successful κ-producing alleles. Importantly, the same heptamer sequence (CACAGTG)
that is the target for the κde rearrangement is the same recombinational
signal that flanks germline V and J regions. This strongly suggests that
the same recombinase that activates this gene (V/J joining) also mediates
its destruction. Rearrangement of the κde eliminates the κ possibility and
may thus ensure λ gene use. In addition the κde may encode a trans-acting
factor that affects λ gene rearrangement or transcription. The κde itself
may mediate the ordered use of L-chain genes and help ensure that B cells
make but one L-chain of a selected class, κ or λ.

In contrast to the B cell malignancies, all T cell neoplasms we have
examined have retained germline κ and λ L-chain genes. Furthermore, most
of the T cell malignancies (23/25) also displayed germline H-chain genes[3].
Similarly, malignancies of nonlymphoid hematopoietic cells including acute
myelogenous leukemia, acute and chronic myeloid phases of chronic myelogenous
leukemia (CML), promyelocytic, monocytic, and histiocyte-like cells all retained germline L-chain genes and usually H-chain genes[3]. Thus, while
H-chain Ig gene rearrangement can occasionally spill over into other lineages, L-chain rearrangements as latter steps in development appear not to.
Therefore, the simultaneous presence of a rearranged H- plus L-chain gene is
a highly B cell restricted marker.

133

NON-T, NON-B ACUTE LYMPHOBLASTIC LEUKEMIAS ARE A DEVELOPMENTAL CASCADE OF B CELL PRECURSORS

The "non-T, non-B" form of acute lymphoblastic leukemia (ALL) was of previously uncertain cellular origin because these cells lacked T cell surface antigens and also failed to display surface Ig. Analysis of their Ig genes proved that while they lacked mature cell surface markers, these leukemias reflected serial stages of B cell precursor development[11,13,14]. All cases examined have shown Ig H-chain gene rearrangements. Approximately 60% of these ALLs retain germline L-chain genes. This provided the first evidence in man that H-chain gene rearrangements precede L-chain gene rearrangements. In 40% of these pre-B cell ALLs which have progressed to L-chain rearrangement, there is also evidence for a κ before λ order to L-chain recombination. We observed seven instances in which the κde had eliminated both κ genes yet λ genes were still germline. Conversely, the five pre-B cells which had moved onto λ gene rearrangement had uniformly already eliminated their κ genes by having earlier rearranged the κde. Thus the Ig gene pattern in these ALLs not only revealed their pre-B cell commitment, but also told us of a developmental cascade of Ig gene rearrangements in which H-chain genes preceded L, and κ rearranged before λ. Moreover, individual leukemic cells are not static, but further developmental progressions have been observed at the Ig gene level between diagnosis and relapse samples[15]. In fact, the earliest of these pre-B cells appear to have D/J intermediate rearrangements which later progress to completed V/D/J. Despite the uniform presence of Ig gene rearrangements the vast majority of these pre-B leukemias fail to make μ chain. Thus, these leukemias may be rich in errors of Ig gene assembly (Fig. 2).

Coordinate with this hierarchy of Ig gene rearrangements, we found a sequential expression of cell surface antigens[13,14]. The very earliest identifiable B cell precursors displayed H-chain gene rearrangement and HLA-DR together with B_4, a B cell restricted antigen reported by Nadler et al.[16]. Later in differentiation, B cell precursors add the common ALL antigen (CALLA) and subsequently rearrange κ or λ L-chain genes. At a variable time of development, the maturing pre-B cells add yet another B cell restricted antigen, B_1[17]. This coordinate sequence of genetic and phenotypic markers has not only advanced our understanding of pre-B cell differentiation, but is providing an important means of subclassifying these leukemias based on their stage of maturational arrest.

ANTIGEN-SPECIFIC RECEPTOR GENES ARE REARRANGING MARKERS FOR T CELL NEOPLASMS

The determination that Ig genes were usually retained in their germline form within most T cells suggested that the antigen-specific receptor present on these T cells was not encoded by the Ig genes. Recently, genes for the β and α chain of the heterodimeric structure comprising the TCR as well as a γ TCR gene have been cloned by the groups of Davis, Mak, and Tonegawa[18-22]. The β-chain gene is the best characterized and is an entirely separate gene complex from Ig, located on chromosome 7. However, it has discontinuous gene subsegments of two constant region genes (C_1 and C_2) and each of these has associated diversity (D_1 and D_2) and joining (J_1 and J_2) regions; a set of variable (V) regions has been identified (Fig. 3). Similar to the Ig genes, the T cell β-chain gene is activated by rearrangements which combine D to J segments or create a complete V/D/J complex.

The β-chain genes (CT_β) must rearrange in any T cell which bears an antigen-specific receptor. This rearrangement, therefore, provides a unique molecular marker for clonal populations of T cells. This is of tremendous importance as there were no routinely available clonal markers for T cell malignancies previously. The rearrangement of the α- and β-chain genes

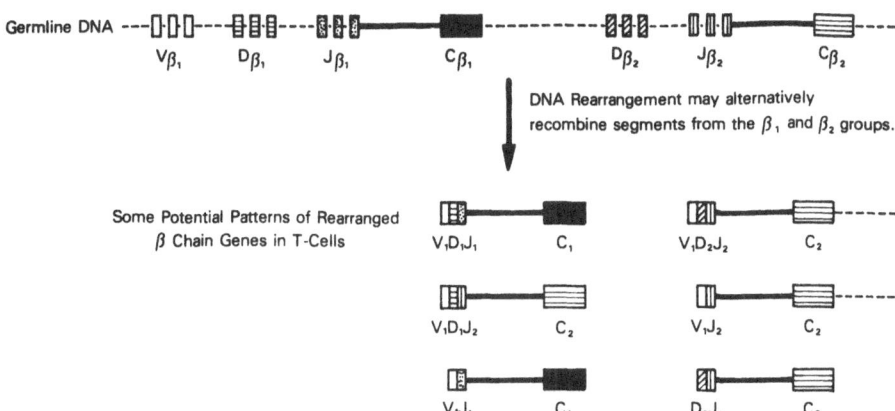

Fig. 3. The human β–chain gene of the heterodimeric T cell antigen receptor has two constant regions (C_1 and C_2) and each has its own joining (J_1 and J_2) and diversity (D_1 and D_2) segments. Variable (V) segments also exist. The spacing of the recombinatorial signals flanking these segments could result in numerous recombinations.

within T cells is permitting us to determine T lineage commitment, clonality, and stage of development of these neoplasms[4].

DETECTION OF CLONAL CELLS WITHIN LESIONS OF MIXED CELLULARITY

The presence of Ig gene rearrangements has been instrumental in revealing that non-T, non-B ALLs and the lymphoid blast crisis phase of CML[22,23] are in reality clonal expansions of B cell precursors. Similarly the leukemic cells of Hairy Cell Leukemia proved to represent a mature B cell stage with the appropriately rearranged, switched, and expressed Ig genes[24]. The application of molecular markers has perhaps made an even greater impact upon our classification and understanding of solid lymphoid neoplasms.

A molecular approach has been crucial in classifying the cellular origin of a number of lymphomas, but particularly those with mixed cellularity. Our attention was directed to several lymphomas which contained a predominance of T cells and small numbers of B cells (Table 1)[3]. Because DNA rearrangements of Ig or β–chain T cell receptor genes are extremely sensitive as well as specific, they can detect clonal populations if they constitute only 1-5% of the total cells present.

Analysis of B and T cell neoplasms has revealed some lineage spill over of H-chain rearrangements in T cell neoplasms (approximately 10%) and TCR β gene rearrangement in B cell neoplasms (approximately 10%). However, κ and λ L-chain rearrangement as a later step in development has to date been solely restricted to B cells.

Table 1 reveals two dramatic instances of lymph node biopsies with a predominance of T cells and a minority of B cells with no cell surface Ig isotype predominance. However, these lesions possessed clonally rearranged H- plus L-chain Ig genes and retained a germline CT_β gene. This molecular

135

Table 1. Monoclonal B Cell Populations Within Some Lymphomas With Predominant T Cells.

	B Cell	T Cell	Surface Ig			Ig Genes T Cell Receptor Gene
	B1	Lyt 3	H-Chain	κ	λ	
Diffuse mixed cell lymphoma	13%	73%	(-)	(-)	(-)	J_H rearranged C_κ rearranged CT_β germline
Follicular and diffuse mixed cell lymphoma	26%	89%	None predominant (G 5%, A 23%, M 28%, D 14%)	(-)	(-)	J_H rearranged C_λ rearranged CT_β germline

analysis established that these are really clonal B cell lymphomas with an infiltrate of polyclonal T cells (Table 1).

The specific rearrangements observed in T or B cell neoplasms provide unique clonal markers specific to these individual neoplasms that enable us to follow their natural history. We have noted developmental progression from one leukemic phase to the next in both ALL as well as in lymphoid blast crises of CML[15,22]. Moreover, the sensitivity and specificity of these clonal markers have enabled us to detect the recurrence of clonal cells within tissue such as bone marrow months before either lymphoblasts or a clinical relapse were noted[15].

CHROMOSOMAL TRANSLOCATIONS ALSO REARRANGE THE Ig GENE LOCI

Specific chromosomal translocations are uniquely or highly associated with histologically distinct neoplasms. Curiously, within mature B cell malignancies the sites of chromosomal translocation involve the very bands that contain the Ig genes. In fact, studies of Burkitt lymphomas have revealed that the molecular site of chromosomal breakage can occur within the Ig gene loci themselves (Table 2)[5,6].

Table 2. Chromosomal Translocations in Lymphoid Neoplasms.

t(8;14) (q24;q32) t(8;22) (q24;q11) t(2;8) (p11;q24)	Burkitt lymphoma
t(11;14) (q13;q32)	Multiple myeloma Chronic lymphocytic leukemia Small cell lymphocytic lymphoma Diffuse large cell lymphoma
t(14;18) (q32;q21)	Follicular small cleaved cell lymphoma Follicular mixed cell lymphoma Follicular large cell lymphoma Diffuse histiocytic lymphoma

In Burkitt lymphoma, the c-myc cellular oncogene locus on chromosome 8 at band q24 is uniformly involved in translocation with the H-chain gene at 14q32, κ at 2p11, or λ at 22q11. The introduction of c-myc into a position near an Ig gene locus markedly alters the regulatory control of this gene.

As Table 2 indicates, the chromosomal breakpoint of 14q32 repeatedly appears within several other mature B cell neoplasms. A translocation between chromosomes 11q13 and 14q32 is observed in multiple myeloma, occasional chronic lymphocytic leukemias, small cell lymphocytic lymphomas, and diffuse large cell lymphomas. In addition, a remarkably common translocation occurs between 18q21 and 14q32 within the very prevalent follicular and some diffuse lymphomas[25],[26]. Thus, the H-chain gene locus appears to frequently mediate chromosomal translocations in human B cell malignancies. Just as c-myc was rearranged in Burkitt lymphomas, there may well be transformation-related genes located at 11q13 and 18q21. As none of the known c-onc genes map to these locations, these chromosomal translocations may be pivotal in identifying new oncogenes.

We have exploited an unexpected rearrangement of an Ig H-chain gene to clone the chromosomal breakpoint in the t(14;18)(q32;q21) chromosomal translocation present in over 60% of human follicular lymphomas[7]. A segment of chromosome 18q21 contains a small 2.8-kb breakpoint cluster region that mediates a t(14;18) break in the vast majority of these translocations. The breakpoints on chromosome 14 were focused at the J_H regions and represent the product of a site-specific recombination that the heptamer and nonamer

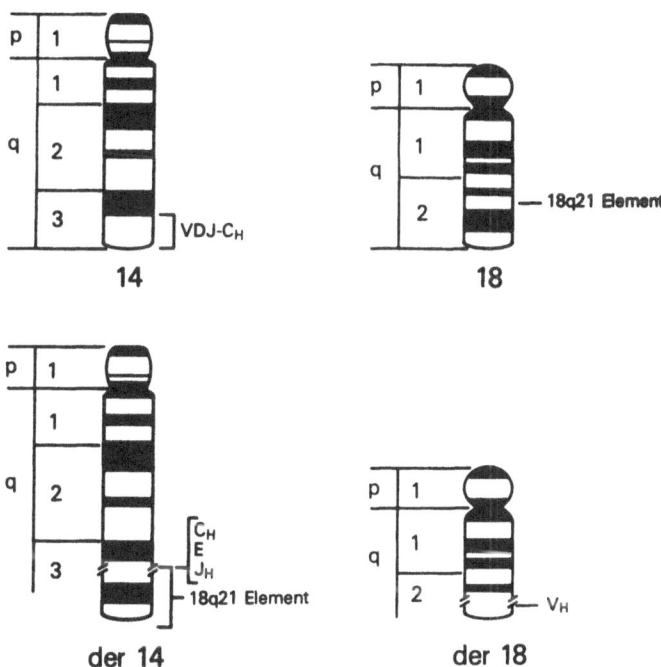

Fig. 4. Classic follicular lymphoma with t(14;18)(q32;q21) translocation. The normal chromosome 14 contains a V/D/J responsible for H-chain production. The derivative (der) 14 chromosome has introduced a new transforming gene site specifically into J_H, near the enhancer (E) element. The reciprocal der 18 partner has received V_H portions of chromosome 14.

signals at the 5' end of J_H appear to direct. A small stretch of extranucleotides not arising from either chromosome 14 or 18 exist at the juncture, compatible with "N" segment addition. This rearrangement introduces a new transcriptional unit from chromosome segment 18q21 into the J_H region in close proximity to the Ig enhancer region (Fig. 4) A 6.5-kb cDNA clone from this locus indicates an elevated transcription of this gene in t(14:18) bearing lymphomas, as well as a lineage restriction of its expression in normal hematopoietic cells. Moreover, the breakpoint cluster region of 18q21 provides a "malignancy-specific" rearrangement that has enabled us to follow the clonal evolution of these neoplasms and to refine the cytogenetic classification of lymphoid neoplasms.

Thus, chromosomal translocations are serving as an additional form of a human gene map. At one side of a chromosomal breakpoint may be a phenotypic landmark gene integral to the counterpart stage of normal differentiation (e.g., Ig in B cells, TCR in T cells). The other chromosome may contribute a transforming gene which when introduced into its new location results in an altered regulation and contributes to the growth or differentiative abnormalities of these tumors.

REFERENCES

1. Aisenberg, A. C. N. Engl. J. Med. 304, 331-336 (1978).
2. Rudders, R. A., Ahl, E. T., Jr. & DeLellis R. A. Cancer 47, 1329-1335 (1981).
3. Arnold, A., Cossman, J., Bakhshi, A., Jaffe, E. S., Waldmann, T. A. & Korsmeyer, S. J. N. Engl. J. Med. 309, 1593-1599 (1983).
4. Waldmann, T. A., Davis, M. M., Bongiovanni, K. F. & Korsmeyer S. J. N. Engl. J. Med. 313, 776-783 (1985).
5. Taub, R. et al. Proc. Natl. Acad. Sci. USA 79, 7837-7842 (1982).
6. Nishikura, K., Ar-Rushdi, A., Erickson, J., Watt, R., Rovera, G. & Croce, C. M. Proc. Natl. Acad. Sci. USA 80, 4822-4827 (1983).
7. Bakhshi, A. et al. Cell 41, 899-906 (1985).
8. Korsmeyer, S. J., Hieter, P. A., Sharrow, S. O., Goldmann, C. K., Leder, P. & Waldmann, T. A. J. Exp. Med. 156, 975-985 (1982).
9. Cleary, M. I., Warnke, R. & Sklar, J. N. Engl. J. Med. 310, 477-482 (1984).
10. Hieter, P. A., Korsmeyer, S. J., Waldmann, T. A. & Leder P. Nature 390, 368-372 (1981).
11. Korsmeyer, S. J., Hieter, P. A., Ravetch, J. V., Poplack, D. G., Waldmann, T. A. & Leder P. Proc. Natl. Acad. Sci. USA 78, 7096-7100 (1981).
12. Siminovitch, K. A., Bakhshi, A., Goldman, P. & Korsmeyer, S. J. Nature 316, 260-262 (1985).
13. Korsmeyer, S. J. et al. J. Clin. Invest. 71, 301-313 (1983).
14. Nadler, L. M. et al. J. Clin. Invest. 72, 332-340 (1984).
15. Wright, J. J. et al. N. Engl. J. Med. (submitted).
16. Nadler, L. M. et al. J. Immunol. 131, 244-250 (1983).
17. Nadler, L. M., Ritz, J., Hardy, R., Pesando, J. M. & Schlossman, S. F. J. Clin. Invest. 67, 134-141 (1981).
18. Yanagi, Y., Yoshikai, Y., Leggett, K., Clark, S. P., Aleksander, I. & Mak, T. W. Nature 308, 145-149 (1984).
19. Hedrick, S. M., Cohen, D. I., Nielson, E. A. & Davis, M. M. Nature 308, 149-153 (1984).
20. Chien, Y.-H., Becker, D. M., Lindsten, T., Okamuras, M., Cohen, D. I. & Davis, M. M. Nature 312, 31-35 (1984).
21. Saito, H., Kranz, D. H., Takagaki, Y., Hayday, A. C., Eisen, H. N. & Tonegawa, S. Nature 312, 36-40 (1984).
22. Bakhshi, A. et al. N. Engl. J. Med. 309, 826-831 (1983).

23. Ford, A. M., Molgaard, H. V., Greaves, M. F. & Gould, H. J. EMBO J. 2,
    997-1001 (1983).
24. Korsmeyer, S. J. et al. Proc. Natl. Acad. Sci. USA 80, 4522-4526 (1983).
25. Rowley, J. D. & Testa, A., Jr. Adv. Cancer Res. 36, 103-148 (1982).
26. Yunis, J. J. Science 221, 227-236 (1983).

HEAVY CHAIN DISEASES

Maxime Seligmann

Laboratory of Immunochemistry and Immunopathology
INSERM U 108, CNRS LP 101
Hôpital Saint-Louis, Paris, France

INTRODUCTION

Human heavy chain diseases (HCD) are lymphoproliferative disorders of B cells characterized by the production of immuno-globulin (Ig) molecules consisting of incomplete heavy chains devoid of light chains.

HCD have been described for the three main Ig classes: γ, α and μ .

HCD raise a number of interesting problems such as the structure of the abnormal Ig molecule and the yet unknown cellular genetic mechanism(s) responsible for the synthesis of a deleted heavy chain and, in most instances, the lack of production of light chains. In addition, the natural history of α HCD, the most frequent condition in the group of HCD, constitutes a model providing unique opportunities for research into the pathogenesis of human lymphoid malignancies.

PROTEIN STRUCTURE

Proteins belonging to each of the four IgG subclasses have been described. When compared to the normal distribution of the IgG subclasses, our data show an abnormally low incidence of γ 2 HCD proteins and a definite increase of γ 3 proteins. More than 100 α HCD proteins that have been typed in our laboratory belonged to the α 1 subclass. The absence of a single case of α 2 HCD protein is certainly not accidental but this finding is presently not understood.

One should recall on one hand that the locus coding for α 2 chain is not close to that coding for α 1 chains. On the other hand, those cases with synthesis of a protein belonging to the α 2 subclass, which could be very rapidly degraded, may perhaps represent non secretory α HCD (see below).

Early immunological and chemical studies have indicated that most HCD proteins consist primarily of the Fc region and have a normal carboxyterminal end. The missing portion of the chain is therefore mainly located in the Fd segment. Sequence data showing that some of these HCD proteins are internally deleted, in vitro biosynthesis experiments and analysis of the mRNA have all demonstrated that HCD proteins are not a mere degradation product of a normal heavy chain. Structural studies (reviewed in 1) have been most complete for γ HCD proteins; they are less numerous for ∝ HCD and μ HCD proteins which have proved more difficult to isolate and sequence.

Contrary to murine myeloma variants, the deleted portion of the HCD proteins never involves only one domain. The deletion usually affects V_H and CH_1 or even three domains (V_H, CH1 and hinge or CH2). The site of resumption of the normal sequence of a heavy chain is not random: the end of the deletion occurs at well defined interdomain regions, often at the beginning of the hinge region. The resumption of a normal sequence is followed by normal constant heavy chain domains, with the exception of a few substitutions or subtle mutations in some proteins.

The findings at the aminoterminal end of HCD proteins show much variation and can be grouped into 3 main categories:

1/ one may find the normal amino terminal sequence of one of the VH subgroups. The length of this normal, incomplete VH stretch varies greatly from protein to protein, ranging from 2 to 100 residues. Such findings establish the existence of an internal deletion that has been recorded for several, but not all HCD proteins. The structural data obtained by Franklin and Frangione (2) for the γ 3 HCD protein Wis. are of particular interest: 3 residues of the aminoterminus are followed by a deletion of most of the VH domain that ends with a small stretch of 8 residues homologous to the J region and this stretch is followed by a second deletion of the CH1 domain ending at the beginning of the hinge.

2/ In several γ HCD proteins, the sequence starts directly within the hinge region. Most such proteins result from a limited proteolytic cleavage of the aminoterminal end following the synthesis of an abnormally short internally deleted heavy chain. Early biosynthesis experiments (3) strongly argued in favor of this view that has been confirmed by the study of the mRNA of the proliferating cells from some patients with such serum HCD proteins (see below).

3/ In some HCD proteins belonging to any of the three heavy chain classes, the aminoterminal stretch preceding the deletion is strikingly abnormal and does not fall into any known sequence of VH subgroups or CH1. The length of this aberrant aminoterminal stretch ranges from a few residues to 42 residues in a μ HCD protein studied in our laboratory (4). The meaning of these aberrant aminoterminal regions is currently unknown. They could result from abnormal splicing, aberrant gene rearrangement or from an insert. In this respect, the possibility of the insertion of an oncogene is of course a tempting hypothesis that should be carefully explored.

MOLECULAR BIOLOGY STUDIES

The molecular biology studies of the proliferating cells are of utmost importance in order to delineate the genetic mechanism(s) responsible for the production of these peculiar HCD proteins. Such studies are still in their infancy and they are actively conducted in our laboratory by P. Guglielmi, A. Tsapis and M. Bentaboulet. The mRNA coding for HCD proteins appears to be abnormally short. Such data, obtained in patients producing HCD proteins belonging to the γ, α and μ classes, exclude the possibility that these proteins result from an extensive post-synthetic degradation or from a mere translational abnormality. In addition, the study of one case of 3 HCD (5) and of one case of α1 HCD (Tsapis and Bentaboulet, unpublished results), with a serum protein starting within the hinge region, showed that the abnormally short mRNA directed the synthesis of a protein larger than the serum protein.

The determination of the sequence of the DNA will allow to decide whether a genomic deletion or a faulty processing of the heterogeneous nuclear RNA or possibly both are responsible for the synthesis of the deleted heavy chains. A genomic deletion encompassing the CH1 domain and its 5' flanking sequences has been demonstrated in three of our cases: one γ HCD (6) and two cases of α 1 HCD (Tsapis and Bentaboulet, unpublished results). In the latter, the deletion also involves the switch region and preliminary observations suggest that the genomic deletion could result in altered or missing splice sites. In contrast, the first results obtained in the study of a case of μ HCD suggest that the production of the abnormal protein results from a mere transcriptional error in initiation or splicing (7). These preliminary results suggest that the mechanisms leading to the synthesis of human HCD proteins may greatly vary from patient to patient. In this respect, data obtained in some murine myeloma mutants producing deleted heavy chains are of interest. For instance, different DNA patterns were found in two α mutants, both producing heavy chains with CH1 deletions of approximately equal extent (8): One had a genomic deletion encompassing all the CH1 exon whereas the other had a smaller deletion that did not remove the entire exon, suggesting that the remaining coding sequences were discarded during RNA processing. The variant of MPC11 that lacks the whole V region provides a precedent for such events since the variable region is removed during RNA splicing (9).

DNA studies in human HCD should involve all intervening sequences in order to investigate sites that are crucial for splicing and to search for inserted material. In this respect, it is essential to perform on the same cells careful cytogenetic studies. Such data are still scarce. Chromosomal abnormalities were found in the lymphoid cells of the mesenteric lymph nodes of three of our patients with α HCD, two of whom had not reached a stage of overt malignant lymphoma (10). In two instances, a rearrangement of 14q32, resulting from a t(9;14)(p11;q32) and a t(2;14)(p11;q32) was observed. One other case showed complex rearrangements including a t(5;9) translocation. No abnormalities were found in the intestinal tumor of a fourth case with immunoblastic lymphoma.

THE LIGHT CHAIN PROBLEM

A main feature of all HCD proteins is the lack of light chains. The nature of the cellular defect leading to this absence of light chains is presently unknown. A failure of light chain production was demonstrated in most cases of γ and α HCD by immunofluorescence and biosynthesis studies (3). However, this is not a constant finding since non secreted monotypic light chains were demonstrated in occasional cases (11). In addition, the proliferating cells from two thirds of patients with μ HCD produce monoclonal κ chains which do not assemble with the abnormally short μ chains and are secreted in urine as Bence Jones protein.

It is of interest that a shortened κ mRNA was found in the proliferating cells in one of our cases of γ HCD (6), as well as in one of our patients with α HCD (A. Tsapis and M. Bentaboulet, unpublished results). We do not know yet if the failure of light chain production is due to some error in light chain splicing and processing or to a structural defect. The link between the genomic defect responsible for the heavy chain deletion and the cellular defect leading to the absence of light chains expression in most cases of IICD has not yet been established. One hypothesis is that the genomic deletion on chromosome 14 could encompass a site crucial for normal processing of κ chains. Another hypothesis is that the light chain abnormality would occur first and select for deleted heavy chains, since the production of entire heavy chains devoid of light chains appears to be a lethal event.

LABORATORY DIAGNOSIS AND CLINICAL PATTERNS (1)

HCD are currently underdiagnosed. The demonstration by immunochemical methods of the presence of Ig heavy chains devoid of light chains in the serum (or other fluids) and/or proliferating cells is absolutely essential to establish the diagnosis and requires a rather specialized clinical laboratory.

When detectable by serum electrophoresis, the pathological protein rarely shows a discrete, localized band or spike suggestive of a monoclonal Ig abnormality since HCD proteins often display electrophoretic heterogeneity. In addition, the pathological protein is not detectable on electrophoretic patterns in approximately 20% of cases of γ HCD, 50% of cases of α HCD and 70% of cases of μ HCD. Although immunofixation may be a helpful procedure, the diagnosis of HCD is usually established by immunoelectrophoretic analysis of the patients' serum, showing an abnormal component reacting with antisera to one class of heavy chains but not with antisera to κ and λ light chains. The lack of precipitation of the anomalous component with antisera to κ and λ light chains may be difficult to ascertain when the concentration of the HCD protein is low. Moreover, it is by no means a sufficient criterion for the diagnosis of α HCD since some whole myeloma proteins, mainly IgA with λ chains, fail to precipitate with most antisera to light chains. In such instances, the diagnosis relies currently on the demonstration of the absence of conformational specificities of the Fab region using highly selected antisera or of the absence of light chain determinants by immunoselection techniques, usually combined with immunoelectrophoresis. The latter method is more sensitive and suitable for testing in routine laboratories (12).

The study of the proliferating cells by conventional immuno-fluorescence or immunoperoxidase techniques is essential for the diagnosis of non-secretory forms of γ or α HCD.

The usual digestive form of α HCD, that involves primarily the small intestine and the mesenteric lymph nodes, represents a true disease entity, whereas the clinical and pathological picture of γ and μ HCD shows much heterogeneity. The striking fact is the disparity between the features of multiple myeloma and those of all types of HCD.

The most frequent presentation of γ HCD is that of a lymphoplasmacytic proliferative disorder, weakness, fever and lymph-adenopathy being the most common initial symptoms. There is no specific histopathologic pattern. It should be emphasized that in close to 10% of the cases, there was no evidence of an underlying lymphoproliferative disorder. Several of these patients were affected with autoimmune disorders and the HCD protein disappeared sponta-neously in some.

Mu HCD appears to be rare. In half of the 16 reported cases, the patients presented with chronic lymphocytic leukemia. Vacuolated plasma cells were found in the bone marrow of at least 10 of the 16 patients.

THE NATURAL HISTORY AND EPIDEMIOLOGY OF α CHAIN DISEASE

Alpha chain disease is by far the most frequent type of HCD. The disease involves primarily the IgA secretory system. It is well established that the lesions progress from an initial stage charac-terized by diffuse and extensive plasma cell infiltration of the lamina propria of the small intestine and of the mesenteric lymph nodes to an overt malignant lymphoma, usually of the large cell immunoblastic type. The lymphomatous cells that arise in the late stage of the disease are derived from the same B cell clone as the initial plasma cell proliferation since they produce incomplete α chains devoid of light chains. Cytogenetic and molecular biology studies should obviously be performed in a given patient at the successive stages of this clonal progression.

That α HCD is not truly malignant in its initial stage is strongly suggested by the occurrence of complete and prolonged remissions achieved in several patients treated with oral antibiotics alone, mainly tetracycline.

The age distribution of α HCD is in sharp contrast with that of multiple myeloma and also of intestinal lymphomas occuring in western Europe, since the great majority of patients were between 10 and 30 years old. The geographic origin of the patients is very peculiar: almost all patients affected with the digestive form of the disease originated from and had been living in areas with a high degree of infestation by intestinal microorganisms: Middle-East, North Africa, Far-East, Indian Subcontinent, Central and South America, Southern Europe. In addition, many of these patients were of low socio-economic background and were exposed to conditions of poor hygiene. These findings strongly suggest that environmental factors, providing a local and protracted antigenic stimulation play an important role in the pathogenesis of α HCD. They could either trigger the clonal proliferation directly or represent only pre-disposing factors. In either case, it is remarkable that the plasma cell proliferation resulting form this postulated antigenic stimulation appears to lead in these patients to the production of HCD protein rather than of a whole myeloma globulin. Systematic microbiological studies, performed in a limited number of cases, have not revealed evidence of a specific intestinal microorganism associated with α HCD and that could be responsible for the first transforming event leading to genomic abnormalities.

CONCLUDING REMARKS

Although the structural analysis of HCD proteins has been completed in a number of cases, the mechanism(s) underlying the synthesis of these abnormal deleted heavy chains and, in most instances, the absence of production of light chains, remain presently unknown. The elucidation of the basic genetic defects in

HCD may help uncover important physiological mechanisms related to Ig gene organization and B cell development. In addition, in view of the aberrant aminoterminal sequences in some of the HCD proteins, molecular biology studies, together with a careful search for chromosomal abnormalities, should aim at the finding of a possible oncogen insertion.

Alpha chain disease constitutes an example of a human lymphoma characterized by a continuous sequence of events ranging from a possibly benign hyperplastic process triggered by intestinal microorganisms and reversible by the administration of antibiotics alone, to an overt neoplastic proliferation. In this context, one should recall some of the findings in Y and μ HCD: the young age of some patients, the occurence of μ HCD in African patients, the frequency of autoimmune disorders in patients with Y HCD, the absence of overt lymphoid malignancies in several patients with this disorder and the spontaneous disappearance of the HCD proteins in some. HCD may thus well constitute important models for the understanding of the pathogenesis of human B cell neoplasia.

REFERENCES

1. Seligmann, M., Mihaesco, E., Preud'homme, J.L., Danon, F. and Brouet, J.C. Immunol. Rev., 48, 145 (1979)

2. Frangione, B., Rosenwasser, E., Prelli, F. and Franklin, E.C. Biochemistry, 19, 4304 (1980).

3. Buxbaum, J.N. and Preud'homme, J.L. J. Immunol., 109, 1131 (1972).

4. Barnikol-Watanabe, S., Mihaesco, E., Mihaesco, C., Barnikol, H.U. and Hilschmann, N. Hoppe Seyler's Z Physiol Chem, 365, 105 (1984).

5. Alexander, A., Steinmetz, M., Barritault, D., Frangione, B., Franklin, E.C., Hood, L., Buxbaum, J.N. Proc. Natl Acad. Sci, 79, 3260 (1982).

6. Guglielmi, P., Bakhshi, A., Mihaesco, E., Brouet, J.C., Waldmann, T.A. and Korsmeyer, S.J. Clinical Research, 32, 348 (1984).

7. Bakhshi, A., Siebenlist, U., Guglielmi, P., Arnold, A., Ravetch, J., Leder, P., Waldmann, T.A. and Korsmeyer, S.J. Clinical Res., 32, 342 (1984)

8. Dackowski, W. and Morrison, S.L. Proc. Natl. Acad. Sci, USA 78, 7091 (1981).

9. Seidmann, J.G., Leder, P. Nature, 286, 779 (1980).

10. Berger, R., Bernheim, A., Tsapis, A., Brouet, J.C. and Seligmann M. Cancer Genetics and cytogenetics (in press).

11. Preud'homme, J.L., Brouet, J.C., Seligmann, M. Clin. Exp.- Immunol, 37, 283 (1979).

12. Doe, W.F., Danon, F. and Seligmann, M. Clin. Exp. Immunol., 36, 189 (1979).

PROLIFERATION AND IDIOTYPE SECRETION BY HUMAN B-CELL NON-HODGKIN'S
LYMPHOMA CELLS IN VITRO

Florry A. Vyth-Dreese and Annemarie Hekman

Division of Immunology, The Netherlands Cancer Institute
Plesmanlaan 121, 1066 CX Amsterdam, The Netherlands

INTRODUCTION

The neoplastic cells in human B-cell non-Hodgkin's lymphoma (B-NHL) can
be regarded as monoclonal expansions of lymphocytes arrested in a particular stage of the normal B-cell differentiation pathway. However, in
recent years it has become evident that this arrest is not absolute and
that neoplastic B cells can be induced to mature further into immunoglobulin (Ig) secreting cells in vitro if provided with B-cell growth
and/or differentiation factors (BCGF, BCDF), which operate in normal B-
cell maturation (reviewed in ref. 1). Recently, interleukin-2 (IL-2) has
been found to exert proliferative and differentiation inducing activity
towards normal and leukemic B cells as well (2,3,4). We have analysed the
functional capabilities of human B-NHL cells upon culture in serumfree
medium without any deliberate addition of growth or differentiation
inducing factors. Proliferation and Ig secretion were found to be related
to phenotypic marker expression, especially of activation-associated
antigens. These parameters were compared with Idiotype$^+$ (Id$^+$) Ig secretion in vivo.

MATERIALS AND METHODS

Cells and Cell Culture Conditions

Five different tumor populations were studied, four of them (MEW, LIV,
KOS and STS) obtained from patients with follicular low grade NHL, one of
them from a patient (TOP) with diffuse intermediate grade NHL. The preparation of mouse monoclonal antibodies (moab) against the Id of the surface Ig (sIg) molecules of patients MEW, LIV, TOP and KOS has been
published previously (5). Tumor samples were obtained from pleural fluid
(MEW), peripheral blood (LIV and TOP), lymph node (KOS) or spleen (STS),
all of these tissues being major affected tumor sites. None of the
patients received chemo- or radiotherapy at the time of tumor sampling.
Tumor cells were isolated by Ficoll/Hypaque density gradient centrifugation and cryopreserved in the presence of 10% dimethylsulphoxide until
use. Cells were thawed in RPMI + 10% FCS (Gibco) (viability >90%) and
immediately transferred to serumfree medium. This medium consisted of
Iscove's modification of Dulbecco's modified Eagle's medium (Gibco)
supplemented with 3.024 g/l NaHCO$_3$ (Merck), ethanolamine (2 x 10^{-5} M,
Merck), bovine serum albumine (BSA, 0.25%), transferrine (35 µg/ml),

insulin (5 µg/ml), linoleic acid, oleic acid and palmitic acid (1 µg/ml) (Sigma) and penicillin and streptomycin (100 U/ml, Gibco) (6). Cells were seeded at 2×10^5 cells per well in 200 µl volumes in flatbottomed microtiter plate wells (Falcon Plastics) and supernatants collected from triplicate cultures after 6-7 days culture, pooled and preserved at -20°C until use. Cells from the same cultures were harvested for measurement of proliferative reactivity or phenotypic marker expression.

Proliferation Assay

Proliferative reactivity was measured by ^{3}H-thymidine (^{3}H-TdR) incorporation after 6-7 days of culture as described previously (7).

Ig Secretions Assay

Id$^+$ Ig secreted by the B-NHL cells into the culture supernatants was measured by enzyme-linked immunosorbent assay (ELISA) as described previously (5) with slight modifications. In brief, goat anti-human IgM (50 µl, 10 µg/ml) was coated on flatbottomed microtiter plates (Costar). After 1 hr at 37°C followed by overnight incubation the plates were filled with phosphate buffered saline (PBS) containing 0.5% BSA, and 30 min later with washing buffer (PBS + 0.1% gelatine + 0.05% Tween 20). In subsequent steps the plates were incubated with 50 µl per well of experimental supernatants (controls: standard Id$^+$ samples and serumfree medium itself), anti-Id (or anti-kappa light chain) moab and goat anti-mouse IgG labeled with horseradish peroxidase. Each incubation was carried out for 30 min at room temperature and plates were washed 5 times in washing buffer between each incubation. As substrate 3.3-5-5'-tetramethyl benzidine (Sigma) was used (stock solution (10 mg/ml in dimethylsulphoxide) directly before use 1 : 100 diluted in 0.1 M Na acetate buffer, pH 6.0 and supplemented with 0.01% (v/v) 30% H_2O_2). The peroxidase reaction was stopped with 0.8 M H_2SO_4 after 10-15 min and absorption measured at 405 nm in a Titertek multiscan (Flow Labs) microtiter plate spectrophotometer.

Immuhofluorescence

Surface markers were determined by direct and indirect immunofluorescence as described previously (7). Kappa or lambda (κ or λ) light chain Ig, IgM, IgD, IgG and IgA were measured with fluorescein-isothyocyanate (FITC) labeled sheep anti-human antibodies; B-cell differentiation/association antigens B1, B2, BA1 and BA2, and Tac antigen (IL-2 receptor) with their respective moab and FITC conjugated goat anti-mouse IgG antibody in a second step.

RESULTS

Phenotypic Characterization

Surface markers expressed on the B-NHL populations are shown in Table 1. No reactivity with anti-T3 or OKM was observed except with STS and TOP populations which were 28% resp. 20% T3 positive. Upon culture in serumfree medium changes were observed in the expression of some surface markers on MEW (decreased IgD, BA2), KOS (increased BA2), LIV (induction IgG, decreased BA2) and TOP (increased IgD, loss of BA1). These changes suggest a slight differentiation towards the plasma cell stage in MEW and LIV populations. However, no full maturation occurred since sIg remained expressed on the cells continuously.

Table 1. Surface marker expression on B-NHL populations

NHL population	Id	IgM	IgD	IgG	IgA	κ	λ	B1	B2	BA1	BA2
MEW	98[a]	95	95	5	−	99	−	95	−	95	55
KOS	100	100	37	−	−	97	−	100	−	96	10
LIV	98	100	−	−	−	100	−	100	2	91	55
STS	70	80	−	42	−	70	−	80	−	43	23
TOP	85	87	50	79	−	85	−	85	5	47	−

[a]percentage positive cells

Proliferative Activity and Id Secretion

Table 2 gives a summary of the functional activities that were measured
after culture of the B-NHL cells in serumfree medium. Also shown is the
expression of Tac (IL-2 receptor) before or after culture. Only MEW cells
(Tac[+]) were found to proliferate "spontaneously" and MEW, KOS and LIV
cells (the latter two acquiring Tac during culture) were found to secrete
relatively large amounts of their respective Id into the culture supernatant. STS cells (Tac[−]) produced very little Id and in cultures of TOP
cells (Tac[−]) no detectable levels of Id were found. These Id levels
secreted in vitro were found to correlate with the Id levels detected in
the sera of the patients (data not shown). Thus Id secretion in vitro
and in vivo seemed to relate to the activation state of the B-NHL cells
and particularly with the expression of Tac.

Table 2. Proliferation and Id secretion related to Tac expression

NHL population	proliferation	Id secretion	Tac expression before / after culture
MEW	15.8[a]	3.67[b]	+ / −
KOS	2.1	0.49	− / +
LIV	0.4	1.60	− / +
STS	0.1	0.05	− / −
TOP	0.9	< 0.01	− / −

[a]cpm ^{3}H-TdR incorporation tested at day 6
[b]µg/ml Id secreted after 6 days culture

DISCUSSION

Our results show that some human B-NHL cells are able to proliferate and
secrete Id positive Ig "spontaneously" in serumfree medium. The capacity
to proliferate in culture was observed only in Tac[+] B-NHL cells. T-deple-
tion experiments (not shown) indicated that T cells contaminating the MEW
culture in low numbers were not responsible for this proliferation. Also
in other B-NHL cultures where low or moderate numbers of T cells were
present, no proliferation was measured after 6 days. Furthermore, the
capacity to produce and secrete Id in culture was restricted to B-NHL
populations that were Tac[+] in vivo or acquired this antigen during
culture in vitro. This induced expression of Tac in culture apparently
was not caused by aspecific stimulation by medium components, and unlikely to be present on T cells, since STS and TOP populations, each
containing considerable numbers of T cells, did not become Tac[+]. Rather
it appears that some B-NHL cells had received signals in vivo by which
they became Tac[+] in vitro. This conclusion is supported by the finding of

a relationship between Tac expression and Id secretion in vitro with serum Id levels in vivo. Only five B-NHL populations were studied so far. Screening of a larger panel of B-cell neoplasms will have to answer the question whether this culture system is suitable to prescreen B-NHL patients for high serum Id levels which counteract the efficacy of therapy with anti-Id antibodies (5). The "spontaneous" proliferation and Id secretion by the B-NHL cells in this study and the related expression of Tac, the IL-2 receptor, suggests that IL-2 itself may be responsible for the observed effects. However, addition of anti-Tac moab to the cultures did not abolish any of the functions of the cells (not shown). Therefore, it is likely that other factors, possibly BCGF or BCDF, are produced by these B-NHL populations. The use of a culture system which to a certain extent reflects the in vivo behavior of B-NHL cells may help in dissecting the role of these B-cell factors in lymphomagenesis (8).

SUMMARY

Tumor cells from 5 B-NHL patients were tested for proliferative activity and capacity to secrete idiotype positive Ig in serumfree medium. A correlation was found between these functional activities and the expression of the Tac antigen (IL-2 receptor) on the B-NHL cells. Moreover, B-NHL patients whose cells secrete Id in vitro, were found to possess high levels of Id in their serum. Therefore, this culture system might provide a means to prescreen B-NHL patients for suitability to treatment with monoclonal anti-Id antibodies.

ACKNOWLEDGEMENTS

We thank Trees Dellemijn for expert technical assistance and Marie Anne van Halem for typing the manuscript.

REFERENCES

1. Kishimoto T, Yoshizaki K, Kimoto M, Okada M, Kuritani T, Kikutani H, Shimizu K, Nakagawa T, Nakagawa N, Miki Y, Kishi H, Fukunaga K, Yoshikubo T & Taga T, Immunol Rev 78:97, 1984.
2. Muraguchi A, Kehrl JH, Londo DL, Volkman DJ, Smith KA & Fauci AS, J Exp Med 161:181, 1985.
3. Lantz O, Grillot-Courvain C, Schmitt C, Fermand J-P & Brouet J-C, J Exp Med 161:1225, 1985.
4. Miedema F & Melief CJM, Immunology Today 6:258, 1985.
5. Rankin EM & Hekman A, Eur J Immunol 14:1119, 1984.
6. Yssel H, De Vries JE, Koken M, Van Blitterswijk W & Spits H, J Immunol Methods 72:219, 1984.
7. Vyth-Dreese FA, Van der Reijden HJ & De Vries JE, Blood 60:1437, 1982.
8. Gordon J, Aman P, Rosen A, Ernberg I, Ehlin-Henriksson B & Klein G, Int J Cancer 35:251, 1985.

NORMAL AND NEOPLASTIC B LYMPHOCYTES PRODUCE INTERLEUKIN-1 (IL-1) AND COLONY STIMULATING FACTOR (CSF)

V. Pistoia, F. Cozzolino, R. Ghio, E. Castigli, M. Torcia, S. Zupo, A. Rubartelli, S. Roncella and M. Ferrarini

From the Servizio di Immunologia Clinica, Istituto Nazionale per la Ricerca sul Cancro, Genova; Istituto di Oncologia Clinica e Sperimentale, Università di Genova; Istituto Scientifico di Medicina Interna, Università di Genova; Istituto di Clinica Medica II, Università di Firenze, Italy

Upon contact with antigen, T cells are activated to proliferate and to release factors that regulate B cell differentiation and also promote an inflammatory process. As discussed in previous chapters of this book, antigen has to be adsorbed and appropriately processed by accessory cells before being presented to T cells in the contest of autologous MHC products.

Encounter with antigen, presented by the accessory cells, is not per se sufficient to induce T cell proliferation. An efficient T cell response requires also the presence of Interleukin 1 (IL1), which induces the synthesis of IL2 by stimulated T cells, thus contributing greatly to the amplification of the response.[1]

The initial studies demonstrated that monocytes/macrophages played a major role as antigen-presenting cells and were also the main IL1 producers. More recently, however, other cell types have been found capable of subserving the function of antigen-presenting cells.[2,3] These data raise the problem of characterizing the cell type responsible for IL1 production in those responses where monocyte-macrophage may not be initially involved.

In other articles of this book, it is demonstrated that B cells can successfully exert the function of antigen-presenting cells. Here we show that B cells can also release IL1.[4] Beside participating in the T cell responses, IL1 mediates also a number of functions relevant to the inflammatory process.[1] This consideration prompted experiments to investigate whether B cells could produce other mediators of the inflammatory process and led to the observation that they release Colony Stimulating Factor (CSF). CSF control the maturation of myeloid precursors and regulate some effector functions of granulocytes and macrophages.[5]

B cells were purified from peripheral blood by removing cells capable of forming rosettes with neuraminidase-treated sheep erythrocytes. B cells were subsequently isolated from the non-T cell fraction by positive selection of cells expressing the B lineage-specific B$_1$ antigen.[4] In the case of tonsil, non-T cells, prepared as above, were treated with the OKM1 monoclonal antibody and complement to remove monocytes and NK cells.[4] These suspensions contained more than 95% B cells and less than 1% monocytes, T cells, or NK cells.

IL1 and CSF production was consistently observed in all of the B cell suspensions tested in the absence of any apparent stimulus (table 1).

Table 1

Production of IL-1 and CSF by normal B cell suspensions

Supernatant:	IL1 activity (U/ml)[2]			CSF activity (colonies/2x10^5 cells)[3]		
	Exp.1	Exp.2	Exp.3	Exp.1	Exp.2	Exp.3
Peripheral blood B cells[1]	129	118	98	23+2	31+4	19+3
Tonsil B cells[1]	115	95	143	19+2	27+2	24+2
Medium (control)	0	0	0	0	3+1	2+1

1. B cells were isolated by indirect rosetting according to the expression of the B1 surface marker and cultured (1x10^6/ml) in RPMI-FCS for 48 hrs.

2. Supernatants were tested at different dilutions using the conventional thymocyte assay. The reported values are relative to those dilutions yielding the highest activity.

3. Supernatants were tested at the 10% final dilution. Target cells for the assay were peripheral blood MNC depleted of E rosetting cells and of adherent cells. Briefly, 2x10^5 cells were cultured in 1 ml of a mixture containing α medium, 10% FCS, 0.3% agar, and supplemented with non-essential aminoacids, L-glutamine, sodium pyruvate. Colonies were counted after 14 days in culture and their cellular composition was assessed by morphological and cytochemical criteria. Results are mean values $\pm$ SE from triplicate cultures.

The hypothesis that B cells were capable of producing IL1 and CSF in the absence of stimuli was reinforced by the observation that B cells cultured in media supplemented with endotoxin-free human albumin also produced IL1 and CSF. These experiments ruled out the possibility that endotoxin in fetal calf serum could stimulate B cells. When B cell culture supernatants were filtered on a Sephadex G-200 column, IL1 and CSF activities were detected in two different fractions of 20-25 Kd and of 65-70 Kd respectively (Fig.1).

Table 2

Production of IL1 and CSF by tonsil B cells fractionated on Percoll density gradient.

Supernatant[1]:	IL1 activity[2] (U/ml)			CSF activity (colonies/2×10^5 cells)[3]		
	Exp.1	Exp.2	Exp.3	Exp.1	Exp.2	Exp.3
Large B cells	75	89	66	33 ± 4	20 ± 2	40 ± 5
Small B cells	22	15	28	6 ± 1	5 ± 0.7	10 ± 1
Medium alone (control)	0	0	0	2 ± 1	1 ± 0.3	0

1. Tonsil MNC were depleted of E rosetting cells and subsequently treated with the OKM1 monoclonal antibody and complement. In order to separate large from small B cells, tonsil non T cells were fractionated on a Percoll density gradient. Four cell fractions were obtained; the first contained low density cells, predominantly constituted by large B cells, whereas the fourth band was highly enriched for small B cells. The second and the third fractions comprised cells of intermediate characteristics. Cells from the first and the fourth fractions only were cultured (1×10^6/ml) in RPMI-FCS for 48 hrs.

2. See legends to table 1.

3. See legends to table 1.

That lymphokine release was a property of activated B cells was confirmed also by experiments (not shown) where small, resting B cells were activated with Staphylococcus Aureus Cowan (SAC) before being tested for lymphokine production. Under these conditions, B cells produced abundant CSF and IL1.

A number of investigators have demonstrated that B cells acquire antigen presenting capacity following activation (see chapters by Parker and Grey).

The present findings showing that also IL1 is released following activation are consistent with those data and add further evidence in favor of the notion that activation renders B cells fully efficient accessory cells. Since activation of B cells is generally induced in vivo by encounter between antigen and specific membrane receptors, it follows that there is an element of specificity determining whether or not a B cell can participate in the induction of immune response. The same appears true for the release of CSF, a cytokine responsible for regulation of myelopoiesis that also controls some steps of the inflammatory process. In this respect, there is considerable analogy between T and B cells in that both cell types release lymphokines following contact with the specific antigen.

The production of IL1 and CSF by the malignant cells from patients with

These data are in line with the molecular weight of IL1 reported by others[6] and also suggest, that CSF produced by B cells may be CSF-1 based upon molecular size.[7] Consistent with the latter notion is also the finding that the large majority of colonies developing in the presence of B cell supernatants contained monocytes/macrophages exclusively (70-80% of colonies). The remaining colonies (15-20%) comprised monocytes/macrophages together with neutrophils. "Spontaneous" lymphokine production could indicate that all B cells were constitutively capable of IL1 or CSF release independently of their state of maturation or activation.

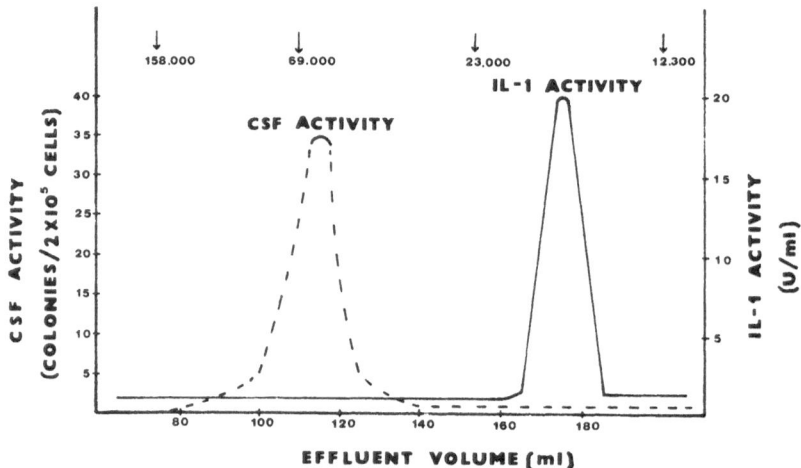

Fig.1 Elution pattern of IL1 and CSF activities produced by normal B cells. Culture supernatants were ammonium sulphate precipitated and were filtered on a Sephadex G-75 column. Fractions were collected and aliquots of each fraction were assayed for IL-1 and CSF activities.

Alternatively, IL1 and CSF could have been produced solely by those B cells that were pre-activated in vivo by a variety of stimuli. The latter hypothesis was explored by fractionating B cells on Percoll density gradients. With this method, large, activated B cells are separated from small, resting lymphocytes. As shown in table 2, CSF and IL1 were released predominantly by activated B cells.

B cell chronic lymphocytic leukemia (B-CLL) was also investigated. As shown in table 3, the cells from 9 out of 12 patients produced both lymphokines.

Table 3

IL1 and CSF production by B cells from patients with B-CLL

Case[1]	IL1 activity[2] (U/ml)	CSF activity (colonies/2×10^5 cells)[3]
A	160	41 ± 4
B	0	5 ± 1 [4]
C	0	7 ± 1 [4]
D	135	35 ± 4
E	42	17 ± 2
F	0	9 ± 1.5
G	93	24 ± 3
H	52	26 ± 3
I	82	21 ± 2.5
J	32	14 ± 1.5
K	21	15 ± 2
L	12	13 ± 1
Medium alone (control)	0	4 ± 1 [4]

1. B-CLL B cells were isolated from the peripheral blood by depletion of E rosetting cells followed by treatment with the OKM1 monoclonal antibody and complement. Purified B cells were cultured in RPMI-FCS for 48 hrs.

2. See legends to table 1.

3. See legends to table 1.

4. Since control values were 4 ± 1 colonies/2×10^5 cells any value below 10 colonies/2×10^5 cells was considered as indicating absence of CSF production.

Previously it was shown that malignant clones from B-CLL patients comprise cells at different stages of maturation and that the proportion of these cells varies in the different patients.[8,9] Thus it is possible that, in those patients where lymphokine production was observed, there was concomitantly an abundant proportion of more mature cells that had acquired the capacity of secreting lymphokines. The observation of "spontaneous" lymphokine production by malignant B cells may also have clinical relevance since these molecules may be responsible for a variety of pathological manifestations observed in these patients.

References

1) Durum, S.K., J.A. Schmidt and J.J. Oppenheim. 1985. Interleukin 1: an immunological perspective. Ann. Rev. Immunol. 3:263.

2) Chesnut, R.W., S.M. Colon and H.M. Grey. 1982. Antigen presentation by normal B cells, B cell tumors and macrophages: functional and biochemical comparison. J. Immunol. 128:1764.

3) Lanzavecchia, A. 1985. Antigen-specific interaction between T and B cells Nature 3134:557.

4) Pistoia, V., F. Cozzolino, A. Rubartelli, M. Torcia, S. Roncella and M. Ferrarini. 1986. In vitro production of interleukin 1 by normal and mali gnant human B lymphocytes. J. Immunol. 136:1693.

5) Metcalf, D. 1984. The granulocyte-macrophage colony stimulating factors. Science 229:16.

6) Matsushima, K.A., Procopio, H. Abe, U. Scala, J.R. Ortaldo and J.J. Oppenheim. 1985. Production of interleukin 1 activity by normal peripheral blood B lymphocytes. J. Immunol. 135:1132.

7) Das, S.K., E.R. Stanley. 1982. Structure-function studies of a colony-stimulating factor (CSF-1). J. Biol. Chem. 257:13679.

8) Fu, S.M., N. Chiorazzi and H.U. Kunkel. 1979. Differentiation capacity and other properties of the leukemic cells of chronic lymphocytic leukemia. Immunol. Rev. 48:23.

9) Rubartelli, A.R., R. Sitia, A. Zicca, C.E. Grossi and M. Ferrarini. 1983 Differentiation of chronic lymphocytic leukemia B cells: correlation between the synthesis and secretion of immunoglobulins and the ultrastruct re of the malignant cells. Blood 62:495.

REFLECTIONS ON THE PHYSIOLOGY OF B LYMPHOCYTES

Benvenuto Pernis

College of Physicians and Surgeons, Columbia University
New York

Reflections on the Physiology of B Lymphocytes

Bone-marrow derived lymphocytes have been said to be the best known cells of the vertebrate body. The purpose of this article is not to review, even in a cursory way, the vast body of knowledge that has accumulated on the function of these cells, but rather to discuss some critical points of their physiology in a way that might help the students of this course in framing their own views of how B lymphocytes react to antigens and ultimately mature into antibody-secreting cells. In particular I shall discuss the relationship of the physiology of B lymphocytes with that of the other main components of the immune system, T lymphocytes and antigen-presenting cells, with an effort to identify the main analogies and differences as well as the biological significance of the functional interactions between these components.

This will be attempted in an informal fashion, with only occasional references to the literature on B lymphocytes, which can be found in recent formal reviews of the function of these cells or in the other articles in this book.

B Lymphocytes and Clonal Selection

Mature B lymphocytes constitute a very diverse population of cells. The basis of this diversity is at the level of the membrane immunoglobulins and, more specifically, of the Fab fragments of these that carry the antigen-binding sites. Numerous germ line genes exist in all species which code for the variable sections of the heavy and light chains of the immunoglobulins. These genes belong to different series (V,D and J for the H chains, V and J for the L chains) that rearrange at the somatic level in immature B cells in the central lymphoid organs (bone-marrow or bursa) generating a first level of diversity that is the result of the particular choice of DNA segments that contribute to the final VH or VL gene. This level of diversity has an additional element at the limits of the D segment due to the possibility of base additions or deletions at those sites. The second level of diversity, that we might call combinational, arises because of the possibility of the B cell to choose between different combinations of a given H chain with one of the many possible L chains or the K or of the λ type. The events that lead to this second level of diversity are, like those that account for the first, strictly limited to the immature B cells in the bone-marrow or in the bursa of Fabricius. The third level of diversity is generated by the possibility of point mutations in the genes that code for the variable regions of the heavy and light immunoglobulin chains. The biochemical basis of the relatively high frequency of somatic point mutations in these genes, limited to B cells and clustered mainly in the complementarity determining segments is not known. It is also not yet definitely established if this third level of diversity is limited to events occurring in immature B cells or if it extends to mature B lymphocytes in the peripheral lymphoid organs. Recent data and views favor the second alternative. If this is so then the third level of diversity provides for the possibility of subtle changes in the specificity of the immunoglobulin molecules synthesized by B cells engaged in a given immune response; selection by antigen of these mutants may provide an explanation of the well known phenomenon of increasing affinity of antibodies during an immune response, with the possibility of reaching very high affinities indeed.

The three levels combined endow the B system with an extremely high degree of diversity of B lymphocytes carrying on their membranes immunoglobulin receptors capable of binding different antigenic determinants (epitopes), that are therefore capable of secreting antibodies of different specificity when they mature to the stage of immunoglobulin-secreting plasmacells. This diversity of B lymphocytes is clearly essential for the production of specific antibodies on the basis of a process of clonal selection, as postulated by Burnet (1) almost thirty years ago.

Another requirement of clonal selection is that, in addition to a large diversity of B lymphocytes at the level of these membrane immunoglobulin receptors, there should be also a strict uniformity of the combining sites of the immunoglobulins present on the membrane of a given cell. The fulfillment of this condition requires a functional hemizygosity of the B cell for what concerns its immunoglobulin genes. This was indeed found to be the case since immunoglobulin genes are subject to the law of allelic exclusion whereby in a given cell and at a given locus only the genes of one chromosome, paternal or maternal, are expressed. The biochemical basis of the phenomenon of allelic exclusion has not yet been settled in detail, but the very fact that it is complete (according to unpublished data from my laboratory F_1 mice have B lymphocytes which show more than 99% exclusion of the two IgD alleles 5a and 5b) strongly favors the possibility that regulatory events, with feed-back from the protein chains, are responsible for allelic exclusion, rather than stochastic events based on a relatively low probability of productive rearrangements of the DNA segments. The same conclusion is supported by the inhibition of rearrangements of L chain genes in transgenic mice that carry one already rearranged copy of such a gene (2).

As the consequence of allelic exclusion of both H and L chain genes and of a strict choice by a given B cell of either one K or one light chain (3), there is strict uniformity of the combining sites of the membrane immunoglobulins present on a given cell (4). This requirement for clonal selection is thus fulfilled by the B cells.

On the other hand clonal selection does not require that all the membrane immunoglobulin molecules synthetized by one B lymphocyte have the same constant region. Most mature B lymphocytes have in fact a mixture of membrane immunoglobulins belonging to two different isotypes, IgM and IgD. These have the same variable regions so that the principle of uniformity of the combining sites of the immunoglobulin receptors of one given B cell is preserved. Alternate splicing of primary transcripts that include the VH as well as the Cμ and Cδ regions explains the capacity of a B lymphocyte to simultaneously synthetize μ and δ chains with identical VH. What is not clear, on the other hand, in spite of

162

extensive research, is the physiological requirement for two classes of receptors on the same cell. It is possible that this requirement has multiple aspects; an interesting and new possibility is that membrane IgM and IgD may have different functions after internalization in endosomes following an interaction with specific antigen. Since this internalization, as is discussed below, has a fundamental role in presentation of antigens by B lymphocytes to T cells, the roles of IgM and IgD in this respect might be different. Here it is pertinent to note that internalized membrane immunoglobulins end up in vesicles that have an acidic pH and are the site of proteolytic activity; under these conditions IgD molecules are split more easily than IgM and this may be of consequence for the ultimate intracellular distribution of the immunoglobulin-bound antigen, or even for the presentation of their own idiotypes by B cells to T cells. It is also interesting to note that in B cells of a certain stage of differentiation there are differences in the capacity to internalize cross-linked membrane IgM in comparison to IgD (5).

In addition to the capacity of simultaneous synthesis of two isotypes B lymphocytes can also switch from the synthesis of one class of immunoglobulins to that of another one, always maintaining the same VH. This is in most cases performed by deletion of an intervening segment of DNA and can take place either in the central or, more frequently, in the peripheral lymphoid organs after stimulation by T-dependent or T independent antigens; in the latter case the switch is predominantly to the constant region genes more proximal to the u/ (IgG$_2$ in the mouse).

Strictly speaking clonal selection is expected to operate through the impact of antigens on an extremely diverse population of B lymphocytes that constitute a random "repertoire" of cells capable of reacting with almost any foreign molecule but lacking clones (the so called forbidden clones) capable of reacting with self constituents. In reality things are quite different. To begin with, although the biochemical events that generate diversity at the DNA level might be expected to operate in a random fashion, in effect this is not so and as a consequence the repertoire of B cells in the central lymphoid organs is quite biased with some VH genes (those closer to the JH region) being more frequently expressed (6).

Even the process that leads to the second level of diversity, that based on the combination of a given H chain with one of several possible L chains, is not entirely random since the combination requires a given level of affinity between the two chains; in fact old observations have established that the disassembled H and L chains of monoclonal immunoglobulins often reassemble preferentially one with another rather than with chains produced by other clones.

In the peripheral lymphoid tissues in turn, there are powerful selective forces, separate from selection by antigens, that deeply influence the repertoire of specificities that antigens will find when they arrive. These forces are those that derive from the possibility of interactions between the secreted immunoglobulins and the lymphocyte receptors; these interactions, also called "network" interactions may be direct ones or be mediated by T cells. As a consequence of these forces some clones or set of clones of B cells that carry a given configuration (idiotype) of membrane immunoglobulins can become dominant for prolonged periods of time and occupy a large section of the B cell repertoire. For instance in a recent experiment (7) mice transfected with a given μ chain gene showed the corresponding idiotype in about one half of the B cell population. Interstingly this was not directly accounted for by the expression of the transfected gene, but was largely due to idiotype regulation that took place not in the bone-marrow but in the peripheral lymphoid system.

Another, and more serious, problem that complicates the function of the B cell system as expected on the basis of simple, direct, clonal selection, is the inability to understand on this basis the distinction of self from non self, that is the fact that the immune system does not normally produce antibodies against the body's own constituents. Actually, B cell clones that can synthesize antibodies against self components normally exist in fair numbers in the repertoire and can be stimulated to produce these antibodies by polyclonal B stimulators. The fact that these clones do not normally produce autoantibodies is probably to be understood through the fact that most if not all potential autoantigens are T-dependent immunogens and that the self-non self discrimination is mainly a feat accomplished by the T system. This simple consideration gives us a reason for the necessity of T cell "help" in order to get antibodies against protein antigens, as is discussed below. In fact experimental tolerance to protein antigens can be induced in the B compartment, but it requires higher antigen dosage and is much less lasting than in the T system (8).

In conclusion the physiology of the B system can be interpreted only partially on the basis of the simple laws of clonal selection, as originally envisaged by Burnet (1).

T-Independent Triggering of B Lymphocytes

There are molecules that trigger B lymphocytes to proliferation and differentiation into immunoglobulin-secreting cells, in absence of any help from T cells. The prototypes of these molecules are the endotoxins of Gram-negative bacteria. These lipopolysaccharides (LPS), of which the biologically active moiety is the lipid (Lipid A), not only stimulate B cells but also are potent stimulators of macrophages which react in a variety of ways such as increased phagocytic capacity, increased synthesis of lysosomal enzymes and much increased production of interleukin-1 (the latter molecule mediates the systemic effects of endotoxins). Other cells, notably T lymphocytes, do not respond to endotoxins in this respect therefore B lymphocytes are more similar to macrophages than to T cells. The biochemical basis of the stimulation of B lymphocytes and macrophages by bacterial lipopolysaccharides is not yet known. A search for specific LPS-binding receptors on the membrane of B lymphocytes has led to inconclusive results; however the capacity to respond to endotoxins, both for B lymphocytes and for macrophages, must be mediated by some defined molecules sinces one autosomal gene appears to control this reactivity: mice of the C3H/Hej strain that are defective in this gene miss both the immunostimulatory and the systemic effects of LPS.

The molecules of the LPS group have one common feature: depending on the concentration they can elicit a specific antibody response (that is they operate as T-independent antigens) or they can activate a large proportion of B cells all the way to immunoglobulin secretion (that is they act as polyclonal B cell activators). Clearly the latter is the basic biochemical effect of LPS, which at lower concentrations can only be exerted on those B lymphocytes that bind LPS molecules through their membrane immunoglobulins and therefore focus sufficient amounts of the stimulating molecule (9). Therefore at lower concentrations a polyclonal B cell stimulator shows its effects essentially as a T-independent antigen.

T-independent antigens do in fact fit in an ideal scheme in which an immune response is based on simple, direct clonal selection of the antibody-producing cells. But then why are not all antigens like that?. Why does the response of B cells to protein antigens require the help of T cells?. I shall touch this problem again later in this article, here I wish to point out that bacterial lipopolysaccharides, as a class of molecules, have structures that do not exist in the components of the vertebrate body (actually a search for endogenous endotoxin-like molecules in various organs invariably ended in the demonstration of bacterial contaminants), so they are clearly "foreign" molecules that do not have to be screened as such by the fine analysis of the T system.

One wonders, therefore, about the evolutionary timing and protective role of the biological reactivity to endotoxins, and in particular of the capacity to produce antibodies to them in a T-independent fashion; one wonders also about the apparent lack of gross immune deficiency in C3H/Hej mice.

In addition to molecules of the lipopolysaccharide group, a variety of other molecules are capable of stimulating B lymphocytes, at least to some extent, with no apparent or only a limited role for help from T cells.

These are, in principle, molecules with a high number of identical antigenic determinants such as different polysaccharides, synthetic polymers or even some polymeric proteins.

The immunogenic capacity of this second category of T-independent antigens is related to their ability to cross-link extensively the membrane immunoglobulins of the relevant B cell clones and push these cells in this way to a given level of stimulation without specific T cell help. Actually specific T cell help is not possible for these molecules since antigen recognition by T cells requires presentation of peptides (in the context of major histocompatibility molecules) that cannot be produced from polysaccharides or synthetic substances like polyvinylpyrrolidone. On the other hand full stimulation of B cells to produce antibodies against this second category of T-independent antigens, which lack the intrinsic stimulatory capacity of endotoxins, requires some levels of T-cell-produced lymphokines, and is therefore not totally T-independent.

B Lymphocytes and Antigen Presentation

T lymphocytes respond to antigenic determinants only if these are part of a protein or are covalently attached to a protein or are covalently attached to a protein (polypeptide) backbone. The reason for this unexpected limitation of reactivity is that T lymphocytes must recognize processed antigen together with molecules of the major histocompatibility complex (MHC).

Although we do not yet fully understand the functional purpose of this limitation of the process of antigen recognition by T lymphocytes, this is nevertheless a well established fact that appears more and more to be valid for different antigen-presenting cells. In principle protein antigens must be processed by intracellular proteolysis and (some of) the resulting peptides presented on the membrane of the antigen-processing cell together with one or another class of MHC molecules. Extracellular protein antigens are internalized by phagocytosis or pinocytosis in macrophages or dendritic cells, split into peptides some of which are then recycled to the cell membrane where the appropriate T cells (in this case normally of the helper group) recognize them together with the Class II MHC molecules that are constitutively expressed on the membranes of dendritic cells or are inducible by the appropriate lymphokines (mostly γ-IFN) on the membranes of macrophages or other cells. It is important to note here that B lymphocytes present protein antigens (that is T-dependent antigens) precisely in the same way. Actually it has been shown (10,11) that antigen presentation by B to T lymphocytes requires internalization of the antigen molecules, limited proteolysis in the endosomes of the B cell and recycling on the surface of the resulting peptides. Since B cells constitutively express on their membrane Class II MHC, this process provides the appropriate molecular set up for the presentation of an antigen to a T cell that has already encountered the same molecular configuration on the membrane of a macrophage or a dendritic cell. Actually the process of antigen presentation again stresses the functional similarities between B cells and macrophages, that have been discussed in the preceding section of this article.

However, there are important differences: B lymphocytes are not phagocytes and they have only limited pinocytosis, on the other hand they possess clonally diversified membrane receptors (the immunoglobulins) that are not synthesized by macrophages or dendritic cells. As a consequence of this condition B lymphocytes internalize and process essentially those protein molecules that bind to their membrane immunoglobulins. Only with very high (non physiological) antigen concentrations B cells become capable of presenting antigens in a non specific fashion, within the range of concentrations of foreign proteins that can be expected in living lymphoid tissues, B lymphocytes are quite specific in their capacity to present antigens to T cells. This specificity is the basis of the specificity of an antibody response to a given protein antigen. What happens in the lymphoid tissues is as follows:

1) Protein antigen is internalized, processed and the resulting peptides presented (with Class II MHC) by ubiquitous dendritic cells or macrophages to T helper cells.

2) Peptide specific T helper cells proliferate and vastly increase in numbers.

3) Rare antigen-specific B cells bind the intact antigen, internalize, process and present it as done by cells in 1).

4) The peptide specific T helper cells that have proliferated as in 2) are numerous enough to achieve a high probability of an encounter with the rare B cells that have bound and processed antigen and now carry on their membranes the appropriate peptide (s).

5) The encounter results in the stimulation of the B cell to proliferate and differentiate into plasmacells that secrete immunoglobulins capable of binding the protein antigen.

This series of events deserves some general comments. First it is interesting that the immune response begins with the dealing of the intact protein antigen by antigen-presenting cells, goes through a phase in which intercellular communication is provided by fragments of the antigen and ends with the prouction by B cells of soluble molecules that again bind the intact antigen.

Second it is clear that our present views have gone a long way from original "antigen focusing" or "antigen bridge" concepts of specific T-B interactions and that we now understand how the necessity of antigen processing by B cells (actually what is processed is a complex of antigen and membrane immunoglobulins) may pose limits to the productive interactions that may occur between different B and T cell clones (the so called phenomenon of the "monogamous" T cell).

Third we may ask: if B cells have antigen-presenting capacity can they substitute for dendritic cells or macrophages in an "in vivo" immune response? This is not so for a simple reason, namely that the antigen binding B cells and the T cells specific for the corresponding peptides are too rare for achieving a sufficient frequency of encounters to start the immune response. Only after the ubiquitous (and non specific) antigen presenting cells have functioned in determining a considerable increase in the number of the T cells capable of recognizing the relevant peptides, is there a significant probability of these encounters. It follows from this last point that, although an antigen-dependent interaction between a B and a T lymphocyte results in bilateral stimulation of the two partners, in essence this interaction has the physiological aim of stimulating the B cell, since for T cells, there is already an overwhelming possibility of stimulation by the members of the macrophage/dendritic-cell family.

Triggering of B lymphocytes by T-dependent Molecules

As I have discussed before most proteins are T-dependent antigens in the sense that production of antibodies against them does not occur in the absence of specific help provided by T cells.

However proteins that bind to the membrane immunoglobulins of B lymphocytes can elicit a series of reactions in these cells even without any intervention of T helpers. These reactions occur at different times after binding of the stimulating molecules to the membrane immunoglobulins and can therefore be distinguished into immediate early and delayed events. The immediate ones, that occur within minutes after the interaction of the ligand, such as depolarization of the cell membrane and influx of Ca ions, are non specific events that follow excitation of many different kinds of cells; they constitute nevertheless an important part of the physiology of the triggering of B lymphocytes. These are followed, after some hours, by a series of early events like a considerable increase in cell size, a nuclear GO to G1 transition, an increase up to 5-6 times of the density of membrane Class II MHC molecules and consequently an increase in the capacity of B lymphocytes to present antigen to T cells. Finally, after one to two days there are the delayed events that consist primarily in a series of biochemical reactions that lead to cell proliferation. What T-dependent antigens (proteins) are not capable of inducing in the absence of T cells or their products is differentiation of B lymphocytes to antibody secreting cells and isotype switches.

A convenient protein ligand for the study of B lymphocyte triggering by T-dependent molecules is provided by allogeneic or xenogeneic anti-immunoglobulin antibodies. These ligands can induce the entire series of T-independent reactions in B lymphocytes provided that they are bi- or multi valent (even better if they are linked to a solid surface) and that they do not engage in the Fc receptors that are on the membranes of these cells. Anti-immunoglobulin antibodies (particularly those of xenogeneic origin) can also function as T-dependent triggering proteins since the B cells can internalize them and present their fragments to T helper cells primed by immunoglobulins of the species that provided the antibodies.

Actually, in the presence of the appropriate T helper cells, anti-immunoglobulin antibodies can work as polyclonal T-dependent stimulators and elicit a generalized response of B cells that includes differentation to plasmacells with immunoglobulin secretion and T-dependent isotype switches.

But, in the presence of the appropriate T helper cells, a stimulation of B cells with the whole spectrum of reactions, is obtained not only with multivalent reagents but also with monovalent ligands' like the Fab fragment (see elsewhere in this volume) that cannot cross-link the membrane immunoglobulins. The interpretation of this surprising finding lies in the fact that B cells do internalize Fab anti immunoglobulin (12) even if much less extensively and more slowly than the cross-linking ligands; recently D. Weber in our laboratory has confirmed this fact by immunofluorescence and has noted interesting coincidence between the internalized Fab and intracellular Class II MHC in B lymphoid cells. B cells therefore show the cellular dynamics that are necessary for the presentation to T cells of foreign proteins that bind to the membrane immunoglobulins even if they do not cross-link these receptors. In fact this is a physiological necessity to provide an antibody response to monovalent protein antigens, like those proteins that are constituted by one chain only, that are not expected to cross-link the membrane immunoglobulin receptors, at least in the absence of piggy-back molecules. The lesson that we learn from the study of the B cell response to anti-immunoglobulin antibodies is that T-dependent (protein) antigens might be divided into two categories: the polymeric ones, or anyhow those with repeated identical determinants on the same molecule, and the monovalent ones.

The first group of proteins can in principle elicit a series of B cell reactions, up to cell proliferation, even in the absence of T cells, but is still strictly T-dependent for the induction of B cell differentiation to antibody-secreting cells. The second group of proteins is completely dependent from T help in the sense that it induces no reaction in B cells in its absence (except perhaps tolerance), but is capable of determining the full spectrum of B cell responses, including differentiation to plasmacells, if T help is available.

The biochemistry of the T - B help is becoming increasingly clear and it appears to be largely mediated by the production by T cells of lymphokines (like the B cell differentiation factor BCDF) that are responsible for the induction of B cell proliferation, maturation to plasmacells and isotype switches. These lymphokines may have an effect limited to a given target B lymphocyte if one considers the possibility that the T helper cell may secrete them locally, in the cleft between its membrane and that of the target B cell, in a way not unlike that whereby specific intercellular stimulation of neurons is provided by the local secretion of pharmacological mediators. On the other hand a T - B help directly mediated through the direct interaction of molecules on the membranes of the two cells (like the Class II MHC on the B cell side) is not yet excluded.

Now I would like to close this rather informal article by raising a question that I believe to be fundamental and is not frequently asked, at least not in an explicit way. The question is: why is it that, in the rules of the immune system, B cells need the help of T cells to produce antibodies to protein antigens?

My view is that B cells do not need help, they need permission. In other words I think that the B cells, with their simple system of immunoglobulin receptors, are not capable (because of cross reactions) to distinguish self proteins from non-self molecules of the same chemical category that might have only small structural differences. T cells on the other hand, in cooperation with the system of antigen-presenting cells and utilizing in a yet mysterious way their restriction to peptides presented together with self MHC molecules, can do precisely that. This might give us a reason for the obvious subordination of the B system to the T system in the immune response to protein antigens.

Acnowledgement: Work in the author's laboratory quoted in this paper was supported by Program Project Grant CA21112-10.

168

REFERENCES

1) Burnet, F.M., 1959. The Clonal Selection Theory of Acquired Immunity, Cambridge University Press, Cambridge, England.

2) Ritchie, K.A., Brinster, R.L. and Storb, U., 1984. Nature, 312, 517-520.

3) Pernis, B. and Chiappino, G., 1964. Immunology, 7:500-506.

4) Raff, M.C., Feldman, M. and de Petris, S., 1973. J.Exp.Med. 137: 1024-1030.

5) Pernis, B. and Roth, P., 1982. Pharmacol.Rev. Vol.34, No.1:65-72.

6) Yancopoulos G.D., Desiderio, S.V., Paskind,M., Kearney, J.F., Baltimore, D. and Alt, F.W., 1984. Nature 311:727-733.

7) Weaver, D., Reis, M.H., Albanese, C., Constantini, F., Baltimore, D., and Imanishi-Kazi, T. 1986. Cell 45:247-259.

8) Weigle, W.O., 1973. Advances Immunol. 16:61-122.

9) Andersson J., Sjoberg, O. and Moller, G. 1972. Transplantation Reviews 11:131-177.

10) Grey, H.M., Colon, S.M. and Chesnut, R.W., 1982. J. Immunol. 129: 2389-2395.

11) Lanzavecchia, A., 1985. Nature 314: 537-539.

12) Metezan P., Eluindi, I. and Goldberg, M.E., 1984. EMBO Journal, 3: 2235-2242.

Sister chromatid exchange, 19, 52
Staphylococcus aureus Cowan I
        (SAC), 42, 67-78, 79, 118,
        121
    activated B cells, 59, 64
Staphylococcus protein A
        (SpA), 67-70
    alternative Ig binding site, 76
Streptococcal group A carbohydrate
        (A-CHO), 45

T cell cloning techniques, 117
T cell lines, 87
TEPC 15, (T15), 84
Thyroglobulin (Tg), 41, 44
T lymphocytes (cells)
    activation, 126
    cloned helper T cell, 84
    helper factors, 110, 114
    hybridomas, 74, 110
    idiotype specific (Th), 84
    T cell leukemia, 123-130
    T cell neoplasm, 133
    T cell produced factors, 25
    T helper, 83-95, 123
    T helper, (Ig specific), 88, 123
    T helper, (Id specific), 88
    T suppressor, 87

T lymphocyte (cell) antigen receptor
(TCR), 89, 123-130, 131
    α chain, 125
    β chain, 125
    TCR genes, 125, 131, 134
TPA, 79, 80
Transformed growth, 53-54